Geoprospection in the Archaeological Landscape

Papers based on contributions to a conference held in January 1989 by The Archaeological Unit, Department of Tourism and Heritage Conservation, Bournemouth Polytechnic

---

edited by

Paul Spoerry

Oxbow Monograph 18
1992

Published by
Oxbow Books
Park End Place, Oxford OX1 1HN

ISBN 0 946897 42 5

Published and distributed by
Oxbow Books, Park End Place, Oxford OX1 1HN
(Phone: 0865–241249; Fax: 0865–794449)

Distributed in the United States of America by
The David Brown Book Company, PO Box 5605, Bloomington, IN 47407
(Phone: 812–331–0266; Fax: 812–331–0277)

Printed in Great Britain by
The Short Run Press, Exeter

Contents

Acknowledgements

This volume could not have been completed without the help and support of many individuals. Foremost amongst these is my wife Pam, whose cartographic, layout, paste-up, proofing and typing skills were invaluable, and whose support and understanding have carried me through a whole host of delays and difficulties.

My thanks must also go to colleagues and friends at the Museum of London Department of Greater London Archaeology and the City of London Polytechnic Department of Geography for providing access to the necessary DTP hardware and software.

I must also thank the referees, and others who have commented on various portions of the text. Arnold Aspinall, John Bintliff, Chris Gaffney, John Gater and Carl Heron all made much-valued contributions of this nature.

My ex-colleagues at the Bournemouth Polytechnic Department of Tourism and Heritage Conservation, specifically John Beavis, Alan Hunt and Bill Putnam gave freely of their time in the organisational stage of the original conference. More recently, access to computing facilities in Bournemouth has again been invaluable.

Finally I would like to thank the individual authors for their contributions, and particularly for their patience. I am also indebted to my co-organiser Steve Thompson for sharing the responsibilities back in 1988/89, and to Arnold Aspinall, who suggested that the conference be held in Bournemouth in the first place.

Paul Spoerry*

July 1991

****Paul Spoerry**, *BTech, PhD, c/o 3 Morley Drive, West Ayton, Scarborough, North Yorks.**

Beyond Wall-Following

Introduction to Parts 1 and 2

This collection of papers is based loosely on the programme of a conference entitled *Geoprospection in the Archaeological Landscape* which was held over three days at the Bournemouth Polytechnic (formerly the Dorset Institute of Higher Education) in January 1989.

The conference was initiated with a number of specific aims in mind. Foremost amongst these was the belief of the organisers that geophysical and geochemical prospection techniques are under-used and under-valued in British (and World) archaeology.

The reasons for this are many. It is probable that an 'image problem' exists; that is, that many archaeologists perceive geoprospection techniques (a catch-all collective phrase for geophysical, geochemical and other, remote-sensing applications[1]) as a quick way of finding walls, so long as the hardware stays in one piece! This comment is not intended to be an insult to 'excavating field archaeologists' as it is really as a result of poor communication on behalf of those of us that use geoprospection techniques in archaeology, that such an outdated (and ill-informed) concept should be in evidence. The various papers in this volume, and many earlier published research projects and surveys illustrate graphically how narrow minded such a view of geoprospection in archaeology really is. But unless direct efforts are made to enlighten and inform fellow non-specialist archaeologists of the great potential for landscape study that is presented by this array of applications, then this volume and any others like it will still only be read by the already converted.

To counteract such a state of affairs the conference included a weekend session (to allow more non-academic archaeologists to attend) which was specifically aimed at selling geoprospection to the mainstream of the archaeological profession. That session is discussed in Part 3 of this volume.

[1] 'Geoprospection' is, of course, an invented word. It is used here to convey a particular meaning, and also as a brief alternative to other more verbose phrases designed to encompass the wide variety of techniques considered in these pages. The particular meaning required of the word in this context precluded the use of 'geoprospecting' as an alternative. The latter, like for example the word 'constructing', indicates the act rather than the science. We are obviously more interested in the science of geoprospection, rather than in describing the act of geoprospecting.

The second explicit aim of the conference was to draw attention to the potential that exists for using geoprospection techniques to understand whole archaeological landscapes, rather than simply to locate, and possibly identify, individual features and/or buildings in isolation. Such concepts reflect ideas that have been discussed in other areas of field archaeology, particularly in field survey (e.g. Cherry 1983, Haselgrove *et al*. 1985). The former author reviewed the type of field survey methodology that should be employed to best understand whole landscapes in both the spatial and temporal dimensions, whilst the latter volume identified the value of studying the ploughsoil assemblage as a cultural continuum across the whole landscape. Also of note is the work of Bintliff and Snodgrass, whose field survey methodology has allowed the interpretation of continuous artefact distributions across the rural *and urban* landscape of ancient Greece (Bintliff and Snodgrass 1985, 1988).

But despite now well-established acadèmic support for the study of field survey material from the whole archaeological landscape, attempts to relate geophysical survey data to such concepts have really only arisen where fortuitous circumstance presented good results from off-site survey adjacent to 'sites' central to the research design. Perhaps the classic example of this is the survey of the prehistoric landscape around the hillfort at Paddock Hill, Thwing, in Yorkshire (Pocock 1982). One exception to this mode of operation is Carr's work in the USA (Carr 1982) where attempts were made to locate American Indian 'activity areas' using resistivity survey. Geochemical survey methods, specifically soil phosphate measurements and soil elemental determinations, plus soil magnetic measurements are tools that are more well-suited to work of this nature, but even these techniques have in the past almost exclusively been applied to specific pre-identified site locations.

This was the theoretical straight-jacket that confronted the conference organisers in 1989. The published literature provided little cause to believe that geophysical techniques were being used in any way that could not be described as glorified wall-following. This is not intended to be insulting to those publications that did indeed present real advances in our understanding of site morphology through geophysics and geochemistry used alongside artefact collection and/or excavation. What was apparent, however, from discussions with those engaged in carrying out geophysical and geochemical surveys was that work *was* in progress in a number of areas that did indeed attempt to go much further than this. The papers assembled here in the first section also entitled *Geoprospection in the Archaeological Landscape* bear testament to this belief. Within this collection of five papers there are examples of archaeological landscapes, complete with their 'sites' which have been studied using an array of techniques that stretches from the archaeological, through the geochemical and geophysical, to the geographical and environmental. The message is undoubtedly clear. The right methods used in combination in the right place can indeed solve problems of real archaeological importance. It is now well accepted that the use of rigorous field survey methodology allows, not only artefact peaks or sites to be identified, but also background distributions to be observed and interpreted and site functions to be postulated. It is now time to place faith in the belief that through applying similarly conceived approaches to geoprospection survey data, a comparable level of understanding can be achieved.

A good illustration of this is afforded by Dockrill and Gater's paper in this volume. Here, not only have geophysical survey techniques helped identify site (mound) type and revealed information concerning the agricultural history of buried soils, but the presence of anthropogenic-related zonation within the survey area has been indicated. This has been achieved through the use of geophysical survey and sampling methods within a wide-ranging project, enabling geophysical results to be tested against excavated data.

Similarly Challands's paper uses magnetic susceptibility data to define site 'edges' and areas of occupation material, cultural indicators which must surely rank in importance at least as highly as a standard 'walls-only' resistivity plot. Bintliff *et al.* describe a set of geophysical and geochemical studies within the wider context of the Boeotia survey. Here, the most valuable information from what are mostly only preliminary studies is that which can be explained with reference to the total landscape collection survey, the approach which forms the core of this multi-disciplinary project. Individual sites are also illustrated, where great success has been achieved in identifying, not only the size, function and morphology of sites, but of different parts of these sites as well; and all this from survey data!

The array of disparate, but complementary techniques used by Passmore *et al.* is undeniably impressive. This study is a prime example of how the physical and chemical properties of both anthropogenic and naturally-derived deposits can provide a generally untapped well of valuable information for those working in environmental archaeology. This successful study begs the question as to why archaeologists of all schools do not spend more energy attempting to understand processes directly relating to the great constant that runs through all our work, that being the soil itself. Recent innovative work, again in the field of 'ploughsoil archaeology', has attempted to understand the variety of processes affecting cultural objects in the topsoil (e.g. Hofman 1986, Entwhistle & Richards 1987, Yorston *et al.* 1990). This perspective is one that needs to be developed further, but it must be considered alongside attempts to understand processes affecting the soil itself. Any attempt to attach significance to archaeologically-derived soil chemical enhancements in survey data must ultimately try to explain the dynamics of movement between the topsoil and subsoil, as features in the latter constitute the reservoir from which it is assumed the enhancement originates.

The second section of this volume is one dedicated to some new techniques of prospection alongside novel approaches to some more established ones. Walker's discussion of the various approaches to soil phosphate measurements that are available is undeniably welcome, whilst Buck *et al.* discuss statistical treatments of soil phosphate survey data, the intention being to identify zones of constant values ultimately as an 'activity area locater'. Bellerby *et al.* describe the novel use of ground temperature measurements as a prospection technique, whilst Noel discusses the potential of multielectrode resistivity tomography as a survey method. The particular value of this last application is undoubtedly as a provider of vertical image sections through deep stratigraphy and ultimately the production of three-dimensional resistivity resolution. At present the only other technique which appears to offer such possibilities in archaeological use, potentially so valuable in urban rescue environments, is survey using ground-penetrating radar (Stove & Addyman 1989).

The third part of the volume entitled 'The Archaeologist and Geoprospection' is discussed at length in a separate section introduction. This title was again that of a session at the conference. The intention was to draw together the practitioners of archaeological geophysics and geochemistry alongside archaeologists from field units, educational establishments and local planning authorities to promote dialogue concerning the potential value of the various prospection methods and to discuss an apparent lack of enthusiasm for geoprospection methods that exists in some areas of the profession.

Over the last three years the role and status of geophysics in British archaeology does appear to have been changing. Two years ago the English Heritage publication 'The Management of Archaeology Projects' (English Heritage 1989) was presented as "a statement of what English Heritage regards as good practice in project management" (*op. cit.*, forward by G.J. Wainwright). Unhappily in the eyes of this editor only one, rather minor, mention of geophysical and geochemical prospection methods was made in this document, that mention being only the inclusion of 'geophysical' and 'soil phosphate' surveys as categories of data to be included, if present, in the site archive. This was despite the first point of the document being that "greater emphasis must be placed on selectivity in excavation and post-excavation work" (*op. cit.*, 3), precisely the area where geophysical surveys are now proving so valuable in many parts of the country as a routine addition to the planning process. English Heritage were not, however, alone in failing to specifically refer to Geoprospection methods in documents in this period. A statement by the ACAO (Association of County Archaeological Officers) in 1988 concerning contract archaeology and competitive tendering had already identified the need for the correct and timely identification in the evaluation stage of any project of the nature and extent of archaeological material under threat. This document, however, still failed to indicate the special role geophysical surveying methods can have in such instances. At about the same time a number of the period societies were also producing documents in response to changes in legislation and an apparent growth in the number and variety of threats experienced by the archaeological heritage. Foremost amongst these was a publication by The Prehistoric Society (1988) which outlined priorities for the protection of prehistoric landscapes under threat. Again, this lengthy document failed to see any specific role for geopropection methods, with only one brief mention of geophysics appearing in its otherwise comprehensive pages.

Such was the background against which the conference session The Archaeologist and Geoprospection was conceived. But despite such an apparent lack of awareness of geoprospection techniques on behalf of so many of the national archaeological bodies in 1988, by 1991 something of a minor revolution appears to have occurred in the use of geophysics as part of the planning process. More and more planning authorities and archaeological units appear to be waking up to the value of properly executed, efficient geophysical surveys, carried out by experienced staff, using modern, reliable hardware and producing interpretations that stand up to excavation. Surveys are now even being commissioned by developers, on advice from planning authorities, as an integral part of the drafting of full planning applications. Perhaps the business acumen of individuals outside of the profession has allowed such concepts to be accepted more easily than amongst the rather conservative ranks of the digging community. None of us are under any illusions that developers would readily fund all necessary archaeology if allowed a free hand, but, where planning constraints are imposed or threatened, it is undoubtedly in the developers interests to carry out an evaluation that is as reliable as possible, as well as one that is relatively cheap. If that means using geophysics, then geophysics will be used.

The almost logarithmic rise in the number of geophysical surveys being conducted as part of the planning process has come about in the absence of any real legislation or national policy as regards our archaeological heritage. It has occurred more through word of mouth and a sort of 'organic' process of development amongst planners, developers and archaeologists. Only in 1990 has any national statement (for England, although similar documents are expected from the Scottish and Welsh Offices) finally been made that acknowledges the value of geoprospection techniques in modern developer-funded archaeology. This is the DoE Planning Policy Guidance Paper on Archaeology and Planning (DoE 1990). This document is undoubtedly of great significance to all who work in archaeology. It has specific relevance to those of us who support the increased use of prospection methods in modern British archaeology as it actually states that developers may wish to carry out geophysical surveys as part of their own initial archaeological assessment, on advice from the planning authority (*ibid.*, section B, para. 20). This may only be one mention of geophysics, but in that specific context it 'paves the way' for geophysical prospection to become a regular consideration in archaeological evaluations, a role to which such techniques are explicitly suited because of their non-destructive nature.

References

ACAO (Association of County Archaeological Officers) 1988, *Contract archaeology and competitive tendering for projects*, Policy Statement.

Bintliff J.L. and Snodgrass A.M., 1985, 'The Cambridge/Bradford Boeotian Expedition: the first four years', *Journal of Field Archaeology* **12**, 123-161.

Bintliff J.L. and Snodgrass A.M., 1988, 'Mediterranean survey and the city', *Antiquity* **62**, 57-71.

Carr C., 1982, *Handbook on soil resistivity surveying*, Central American Archaeological Press, Evanston, Illinois.

Cherry J.F., 1983, 'Frogs around the pond: Perspectives on current archaeological survey projects in the mediterranean region', in Keller & Rupp (Eds.) *Archaeological Survey in the Mediterranean Area*, BAR (S) **155**, 375-416.

DoE, 1990 (Nov), *Planning Policy Guidance 16: Archaeology and Planning*, HMSO.

English Heritage, 1989, *The Management of Archaeology Projects*, English Heritage.

Entwhistle R. & Richards J., 1987, 'The geochemical and geophysical properties of lithic scatters', in A.G. Brown & M.R. Edmonds (Eds.) *Lithic Analysis and Later British Prehistory*, BAR **162**, 19-38.

Haselgrove C., Millet M. and Smith I., 1985, *Archaeology from the Ploughsoil: Studies in the Collection and Interpretation of Field Survey Data*, Department of Archaeology and Prehistory, University of Sheffield.

Hofman J.L., 1986, 'Vertical Movement of Artefacts in Alluvial and Stratified Deposits', *Current Anthropology* **27**, 163-171.

Prehistoric Society, 1988, *Saving Our Prehistoric Heritage: Landscapes Under Threat*, Report by the Prehistoric Society.

Stove G.C. and Addyman P.V., 1989, 'Ground probing impulse radar: an experiment in archaeological remote sensing in York', *Antiquity* **63**, 337-342.

Yorston R.M., Gaffney V.L. and Reynolds P.J., 1990, 'Simulation of Artefact Movement due to Cultivation', *Journal of Archaeological Science* **17**, 67-83.

Part I

Geoprospection in the Archaeological Landscape

Trace Metal Accumulations in Soils on and around Ancient Settlements in Greece

J.L. Bintliff, B. Davies*, C. Gaffney, A. Snodgrass• and A. Waters••**

*Department of Archaeology, University of Durham, *Department of Environmental Sciences, University of Bradford, **Geophysical Surveys of Bradford,•Museum of Classical Archaeology, University of Cambridge, ••Department of Archaeological Sciences, University of Bradford.*

Abstract

This paper describes recent studies of trace metal and magnetic enhancements in soils on and around ancient settlements in Boeotia, central Greece. The investigations were conducted as part of the Cambridge and Bradford Boeotian Archaeological and Geographical Expedition and the scientific data were collected with reference to archaeological sites originally identified as artefact scatters. Transects of soil samples were taken across the whole area under study to provide context for the site-related data. Elemental analysis of samples was conducted through AAS. The results indicate enhancement of lead, zinc and copper, but not nickel, in soils around known ancient sites. Detailed analysis of trace metal concentrations and magnetic enhancements, in relation to artefact distributions and geophysical data, provides indications of site function and morphology in several instances. The trace metals data also supports artefact collection data off-site, in areas where intense infield activity is proposed.

Introduction

Modern evidence suggests that wherever people live or work metal concentrations in nearby soils rise, the metals are bound in chemical forms which are often not susceptible to diminution by leaching, and the resultant accumulations can still be detected several centuries later. This suggests the possibility that much more ancient settlement could also be associated with unusual and localised accumulations of certain metals in soil. Other soil properties, notably soluble phosphorus content and magnetic susceptibility, are commonly studied as an aid in archaeological research, together with other changes in soil properties revealed by cropmarks visible in air photographs. The success of these methods suggests that changes which are imposed on the soil mantle can be identified many hundreds of years, even millennia later.

The objective of a recent investigation was to determine whether certain pre-Industrial archaeological sites in Greece were characterised by unusual accumulations of trace metals in soil. Several metals were selected. Copper and lead compounds have been used since later prehistory as has zinc when alloyed with copper (brass), even though metallic zinc was not known in Europe until the 15th century. Nickel is a modern metal and there is no reason to suppose it should have accumulated in ancient sites other than around copper ore workings. Manganese was used in the ancient world to bleach glass but the metal was not isolated until the 19th century; its soil chemistry suggests it is not a good candidate for an archaeological marker but it is a useful indicator of pedological conditions. However all of these elements may concentrate in animal and human faeces.

The samples were collected during the 1986 and 1987 field seasons of the Bradford and Cambridge Boeotian Expedition. Since 1979 the Boeotia Survey has systematically fieldwalked over 40 sq.km. of countryside in the central Greek province of Boeotia (Bintliff and Snodgrass 1988a). Ancient habitation sites have been identified from concentrations of surface artefacts (essentially pottery and tile) and by mapping 'offsite' background densities of ancient artefacts between these sites in order to trace past human activity such as field manuring and work stations in the cultivated landscape (see the sector mapped in Figure 1 with offsite density variations). It should be pointed out that our geomorphological researches suggest that the vast majority of this landscape has not been buried by eroded material since the time of the sites being considered in this paper, ie Hellenistic times; however it has generally suffered topsoil erosion into localised lowlying depressions. In other words we consider this to be a truncated relict landscape.

Regional Trace Metal Survey

In order to provide regional baseline data for trace metal analysis and to seek evidence for any long-distance trends, surface soil samples were collected along several long transects (Figure 2). Two east-west lines of transect were established, 500m apart but in parallel, and both ran for 4km west from the edge of the ancient city of Thespiae. At right-angles two further transects, running north-south, were set up, running for 2 and 5km respectively. Samples were taken at 200m intervals.

On Site Trace Metal Survey

A series of surface sites was chosen for each of which a grid of soil samples was taken for trace metal analysis.

They included the ancient city of Thespiae, the medieval village at VM4, and Greco-Roman rural villa sites at TPW2, PP17, PP27, VM64, VM89 and VM95. In most cases complementary information was available from geophysics (for structural evidence), magnetic susceptibility (reflecting habitation refuse and pronounced soil disturbance) and surface artefact patterning (indicating living, working and rubbish disposal areas).

Field and Laboratory Methods

Samples were collected using either a mild screw auger or a stainless steel garden trowel. After drying and chemical treatment metal contents were determined using conventional flame atomic absorption spectrophotometry.

Summary Results

Inspection of the results (Figure 3) indicates a general tendency for the lead, zinc and copper values from the specific archaeological sites to be greater than those from the regional transects, whereas the manganese values are smaller except at PP17. Nickel values vary widely above and below the regional mean.

There was no evidence for consistent accumulations of nickel at the archaeological sites and this is consistent with the initial hypothesis that pollution by this metal is essentially a modern phenomenon. As for manganese, the regional mean of 761 mg/kg is consistent with soil averages in the literature of c.1000. Except for PP17, site values are significantly below the regional mean. Whether the anomaly at farm PP17 could be the result of stockpiling manure remains to be investigated. The zinc values are variable, with the means at Thespiae and VM4 significantly higher than the regional mean, whereas at farm sites there can be both abnormally high as well as normal background values. This suggests that zinc may accumulate at village or urban sites where metal working and other industrial activities undoubtedly took place, but also and for unexplained reasons at many rural sites.

Copper and lead were identified at the start of the investigation as two of the elements most likely to be markers of vanished human occupance of the land. Davies and others (cf. Davies 1978) have demonstrated the ubiquity of soil contamination by copper and lead throughout the British Isles in garden soils of both city and country. For pre-Industrial times, an obvious source for this 'habitation effect' are metallic compounds used since later prehistory for coins, ornaments, glazes and pigments. However equally and possibly more relevant in rural archaeological contexts is the evidence from Classical written sources and pottery dispersal on and off site (cf.Bintliff and Snodgrass 1988b), that long-occupied agricultural landscapes and ancient farm and village sites may have zones of accumulation of human and animal waste products, especially in the form of manure, which may contain abnormal concentrations of these and other trace elements. Hence the hypothesis that soils associated with ancient habitation would contain residual accumulations of these metals. Only at site PP27 (where in any case we had low sample numbers), was there no evidence of an excess of copper over the regional background; lead was always elevated at the sites sampled.

Perhaps the most dramatic illustration of these effects can be seen at the ancient city of

Thespiae, where the sampling transect ran from the countryside across the ancient wall into the former town enclosure (Figure 4). Here metal values soar inside the walled area.

Detailed Analysis of Rural Farm/Villa Sites

However the potential of the approach and its problems are best seen in smaller rural sites, typical for field survey, and where the activities involved anciently are much less understood compared with urban or large village sites.

PP17

This site was located through surface finds of roof tiles and potsherds and identified as a small farmstead of Late Hellenistic to Early Roman times. Possible outlines of the collapsed farmhouse came from resistivity survey (Figure 5), together with additional features provisionally interpreted as farm enclosures and perhaps pits beyond. Contouring of the surface pottery (Figure 6) showed the highest concentration left of the farmhouse, within and beyond the suspected yard; the rooftile however was predictably focussed on the dwelling structure. Interpretation here as elsewhere suggests that domestic rubbish disposal is associated with but not peaking within the living structure.

The trace metal sample grid covered a much larger area beyond these archaeological and geophysical features defined as 'site' (Figure 7). The lead plot (Figure 8) and the copper (Figure 9) show values, even so, all well above the regional norm; all values shown are in excess of background (eg minimum lead is 22mg/kg compared to the regional norm of 6.6, copper minimum is 8.4 compared to 5.7). Clearly the trace metals are picking up a wider area of past human activity than the habitation zone proper. Indeed a further set of samples taken in 1987 found that high values continue for both metals tens of metres beyond these diagrams away from the 'site'. Now if we look at PP17 in terms of offsite pottery densities (Figure 10) we can see at once that there is indeed what we call a strong site halo of high discard extending at least 100m in all directions from the site focus. We interpret the combined evidence as showing intensive infield activity based at the farm, probably combining concentrated manuring, rubbish disposal and localisation of farm animals.

Further insights can be obtained if we look at a close-up of the site proper. The copper plot (Figure 11) shows a discrete high over the query two-roomed farmhouse, with a second major concentration on the left of the picture; the lead in clear contrast (Figure 12) forms a ring of high values around the farm but the actual structure is a pronounced trough of values. We seem to be picking up differential accumulation of the two metals across the site and in its halo, which should eventually shed light upon behavioural variations on this relatively short-lived site.

Apart from building-up a series of case-studies of farmsites like PP17 to improve our understanding of such phenomena, a further avenue we hope to explore is taking comparable suites of samples from the extraordinarily well-preserved Classical farmsites in nearby Attica (Lohmann 1983, 1985), where the structural evidence is still free-standing together with contemporary field systems and agricultural installations.

VM64

This is another small farm site (Figure 13), this time of Imperial Roman date. The pottery contours identify a neat concentration with a northward tongue, and a fringe area to the south on a higher terrace. The tile counts (Figure 14) identify essentially the same habitation focus, with a slight shift in peak values compared with the pottery peaks, yet also a northward tongue. For unexplained reasons at this site domestic discard broadly coincides with what seems to be the collapsed farmhouse. The geophysical interpretation (using an advanced Schlumberger array) (Figure 15), compared with the tile and pot counts may be picking up one massive end and perhaps another boundary of this farmhouse, but also further unidentified features in the very top left and right sectors. Immediately adjacent to the query house is what could be an untiled yard or shed.

Magnetic viscosity measurements (Figure 16) echo the rooftile closely, and the mysterious northward tongue. Magnetic susceptibility (Figure 17) has an almost identical distribution. So far then, the local separation seen at PP17, of main structure with tile from the main concentrations of refuse and human activity traces in the soil, is contrasted at VM64 where we see a closer association. This contrast continues with the trace metal data.

Lead for example (Figure 18) has a strong concentration over the main structure at this site, although we note immediately that as at PP17 high values appear in the upper sectors where only the geophysics had shown activity. As for copper (Figure 19) as at PP17 one peak sits discrete over the farmhouse, whilst the other peaks are in more peripheral sectors, often where other indicators apart from geophysics are absent. The northward tongue is a trace metal low in both cases.

Just as at PP17 these hints from the trace metals of a wider radius of past human activity can be followed into a wider halo, beyond the ostensible archaeological site with its structures and artefactual concentrations (Figure 20). Four transects were set up running 50 metres in the cardinal directions away from the site focus. Along each, soil samples were taken for magnetic susceptibility and trace metals. The South-North pair (Fig. 21) for magnetic susceptibility is here placed against the tile counts across the site proper; we see a good agreement within the site, yet the highest magnetic values are just offsite to the south. Likewise with the East-West pair of transects (Figure 22), a very nice match of tile counts and magnetic susceptibility occurs on site, but the largest peak is just off site.

The trace metal results are even more striking: for copper (Fig. 23a) we can note that all values are well above the regional mean of 5.7, but the site focus peaks are matched and exceeded by peaks around the site. Likewise for lead (Fig. 23b), and here we do see a closer parallel to PP17 with the site core as a relative trough compared to the immediate offsite zones. Note again that all values are above the regional mean of 6.6. Once again if we take a bird's eye view of the offsite pottery map (Figure 24) we see confirmation that VM64 has a well-developed discard halo, corresponding to heightened activity around the formal site being indicated by magnetic components and the trace metals.

TPW2

This is a much larger villa, typical for the Late Roman period. As may be seen from the regional sherd density map (Figure 25) this 3.4 hectare site lies in an area of extensive halo effects, in which its own substantial halo is merging with the gigantic halo effect of discard from the city of Thespiae nearby to the right of the picture.

Detailed analysis has only been made of a small hillock at the heart of the site. The tile count data for this sector of 60 by 40 metres (Figure 26) bring out two likely roofed structures and a lack of significant roofing to the left of the plot. The magnetic viscosity plot (Figure 27) picks out the same two peaks but introduces a lesser accumulation in the far left. As for magnetic susceptibility (Figure 28) we merely see the overall tendency mirroring the tile, to emphasize the right hand sectors versus the left on the sampled area. The resistivity plot (Figure 29) is only available for the bottom limb of the L shape, but does give strong results: a massive east-west wall, a lesser north-south wall crossing it and what may be the corner of a structure in the upper right. If we overlay geophysics with tile (Figure 30) it does look as if one roofed structure is defined by the lower enclosing walls, and the other roofed structure, seemingly larger, is defined by two walls running at right angles. It is quite likely that the isolated wall stump on the right marks one corner of the upper farm building. The upper left sector seems well defined as a separate enclosure, with just a thin trail of tile above the wall maybe suggesting a shed or similar feature built up against the wall; interestingly the magnetic viscosity plot picks out this query feature as well.

Now for the trace metals. The copper plot (Figure 31) picks out the lower roofed structure with a small peak, and larger ones for the upper roofed structure. But we also have a substantial accumulation in the neglected enclosure to the upper left. The lead plot (Figure 32) is peaking well over the lower structure, but peaks sit peripherally to the upper right structure and there is again accumulation in the upper left enclosure.

Comparing TPW2 with our other rural sites we see once more that copper picks out the roofed structures, whereas lead overlies some structures whilst accumulating around others. Both metals are also accumulating in parts of the site lacking obvious structural evidence or artefact peaks. Within the site core the remnant magnetism associates closely with the roofed structures, but can be expected to appear in additional peaks peripheral to the habitation features identified by archaeology and geophysics.

Conclusions

We clearly have been able to demonstrate our initial hypothesis, that trace metals accumulate at very significant excess levels on and around ancient sites (even if as at PP17 occupation may have lasted a mere 200 years), and furthermore that the patterning persists up to the present day. But is this going to be yet another example of the 'archaeometry paradox', where we produce scientifically satisfying analyses of archaeological data which nonetheless neither surprise the archaeologist, nor lead on to solve any important archaeological problems? No, in this case we would like to indicate some substantial implications of the research results just outlined:

(1) Firstly it has become clear that our concept of the 'site' should envisage on the one hand the traditional focus with buildings and peaks of artefactual refuse, but in addition a surrounding halo or infield of intense human activity, showing up from plateaux of offsite pottery discard and accumulations of magnetism and copper and lead that are often above site focus levels. Both zones have values all well in excess of the regional norm. The research frontier here will be to model ancient behaviour, and try to identify recurring suites of patterning, as well as to investigate accumulation patterns in modern farm contexts.

Site survey is a major archaeological tool. Only a tiny fraction of sites discovered will ever be excavated, so the application of this kind of battery of subsurface and surface approaches is a vital aid to interpretation of field survey data. In fact it is probable that excavation in the halo area would in any case find little or no structural evidence.

(2) Secondly, these results shed unexpected light on a major issue in landscape archaeology, the effect of soil erosion on the rise and fall of past complex societies. For Greece Van Andel (cf.Pope and Van Andel 1984) and other geomorphologists have shown that the cyclical collapse of prehistoric and Greco-Roman settlement can be associated with massive episodes of soil erosion that presumably ruined the agricultural economy. Our own work on relative densities of surface pottery (Bintliff and Snodgrass 1988b) shows a cline of ever greater amounts of surface pottery as one moves from England, through the Mediterranean, to Arabia, which we have interpreted as reflecting amongst other processes a cline of increased soil erosion.

Yet our Boeotian data allow us to set limits to the scale of these erosion processes. The pottery and tile peaks over site foci can of course remain even if the soil fines, their original context, have washed away, but the peak accumulations of remnant magnetic components and copper and lead within site foci and in the immediate halo can only be explained through the survival in situ of at least the original subsoil of the ancient sites. In support of this view Professor Van Andel has estimated that the total depth of soil lost on average in the last 5000 years in the southern Greek landscape is less than 1 metre (Van Andel et al.1986, p.111).

(3) The trace metals programme may well prove vital in shedding new light on past intensity of land use across the countryside. Already we have used the offsite pottery densities to suggest (cf.Bintliff 1988) that although field survey never finds all the ancient farm sites, the low numbers in the north of our survey area genuinely reflect less intensive land use. It should be possible (and we are already experimenting with area grid soil collection strategies in Yugoslavia and Greece to this purpose), to match trace metal accumulations with offsite pottery discard, if we are correct in our belief that much of the excess copper and lead reflect human and animal manure spread over the cultivated landscape.

Quite unintentionally our Boeotian regional trace metal transects, although designed as a control over anthropogenic accumulations at habitation sites, hint at the potential of offsite trace metal analysis for land use history. We must point out that the sample collection design could not have been less suitable for obtaining a picture of local manuring patterns, with tiny soil samples gathered at 200m intervals. Even so, if we note how the east-west transects run from lowish offsite pottery discard into higher levels nearing the city of Thespiae, undoubtedly reflecting urban infield manuring, so on the mean plots for copper and lead (Figures 33 and 34) we see a general trend of rising values towards the city. We can even say something about some of the peaks or hollows on these graphs, covering as they do 4 km of landscape: quite a few of the localised peaks mark sectors where soil samples were taken within site haloes (Figures 35 and 36).

Clearly we need to take many more closely-spaced soil samples for an accurate comparison with field-by-field offsite pottery density values, in order to test the usefulness of trace metal assay for past land use intensity. Trace metals should be much more firmly attached to the soil matrix than phosphates and their analysis could act as a complement and in many soils a replacement for phosphate analysis at the landscape level.

References

Bintliff J.L., 1988, 'Site patterning', In J.L. Bintliff, E. Grant & D. Davidson (eds.) *Conceptual Issues in Environmental Archaeology*, 129-144. Edinburgh, Edinburgh University Press.

Bintliff J.L. and Snodgrass A.M., 1988a, 'Mediterranean survey and the city', *Antiquity* **62**, 57-71.

Bintliff J.L. and Snodgrass A.M., 1988b, 'Offsite-pottery distributions', *Current Anthropology* **29**, 506-513.

Davies B.E., 1978, 'Plant available lead and other metals in British garden soils', *Science of the Total Environment* **9**, 243-262.

Lohmann H., 1983, 'Atene:eine attische Landgemeinde klassischer Zeit', *Hellenika Jahrbuch*, 98-117.

Lohmann H., 1985, 'Landleben im klassischen Attika', *Jahrbuch Ruhr-Universitat Bochum*, 71-96.

Pope K.O. and Van Andel T.H., 1984, 'Late Quaternary alluviation and soil formation in the Southern Argolid', *Journal of Archaeological Science* **11**, 281-306.

Van Andel T.H., Runnels C.N. and Pope K.O., 1986, 'Five thousand years of land use and abuse in the Southern Argolid, Greece', *Hesperia* **55**, 103-128.

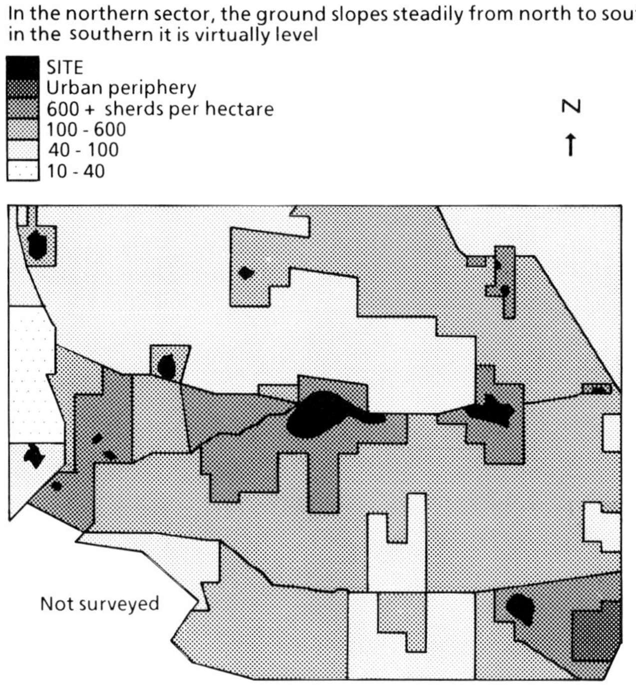

Figure 1 *A typical Boeotian density plot.*

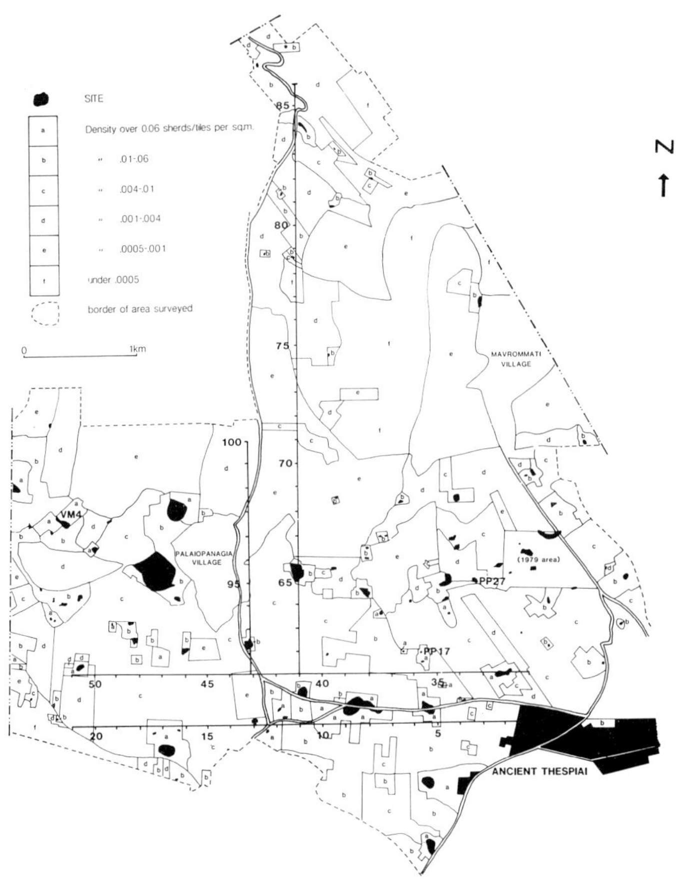

Figure 2 *Regional trace metal survey transects.*

	METAL				
	Pb	Zn	Cu	Mn	Ni
	SITE MEAN SOIL METAL (mg/kg)				
REGION	6.6	6.6	5.7	761	192
THESPIAI	13*	18***	19***	239***	113
PP17	53***	7.5	13***	1019***	69***
PP27	11*	5.0	6.7	171***	89***
TPW2	23***	49***	23***	478***	446***
VM4	20***	17***	21***	70***	89***
VM64	16***	65***	26***	624*	232
VM89	19***	55***	28***	604***	473***
VM95	15***	55***	21***	532***	254

Values different from the regional mean are shown at the following significant levels: *** 0.1%, ** 1%, * 5%.

Figure 3 Mean concentrations of metals in soils at individual sites and of the region as a whole.

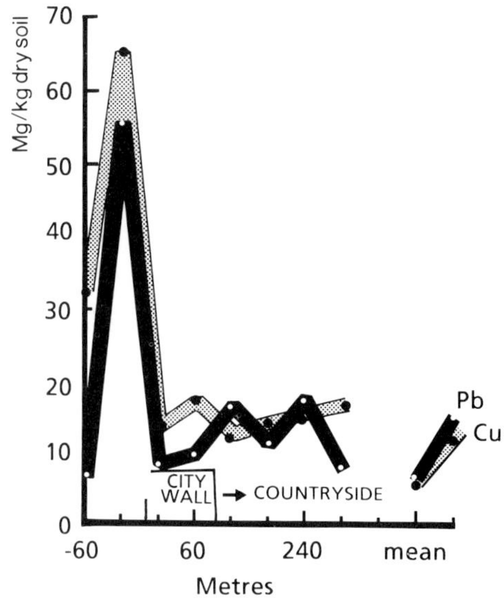

Figure 4 Copper and lead concentrations in soil samples taken along a transect crossing the ancient city walls at Thespiai.

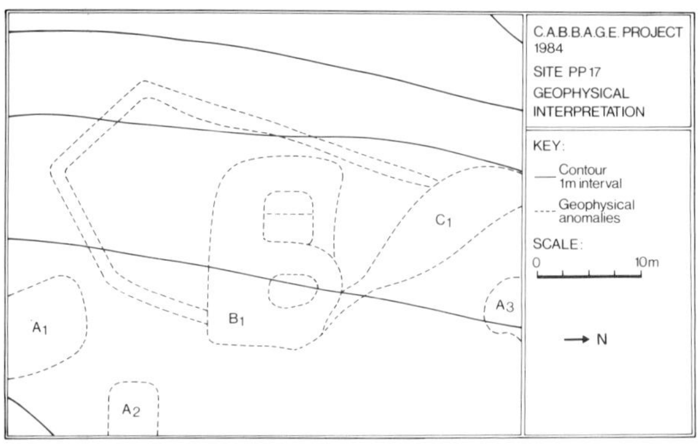

Figure 5 Interpretation of resistivity survey anomalies at PP17.

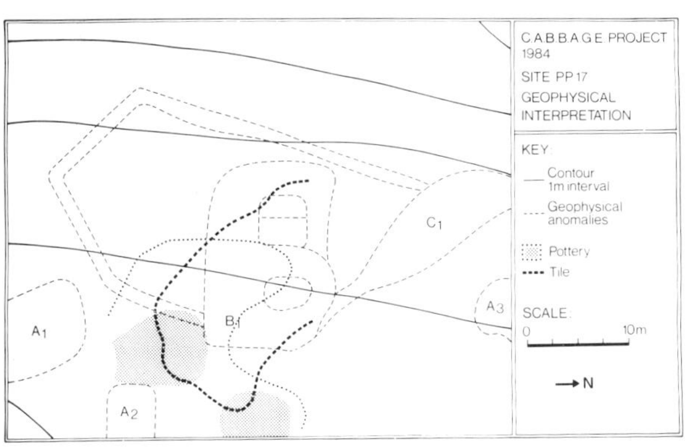

Figure 6 Location of tile and pottery concentrations at PP17.

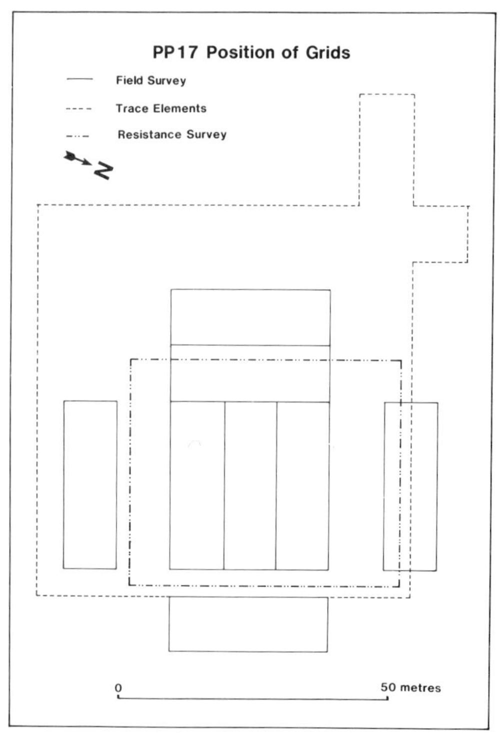

Figure 7 Location of survey grids at PP17.

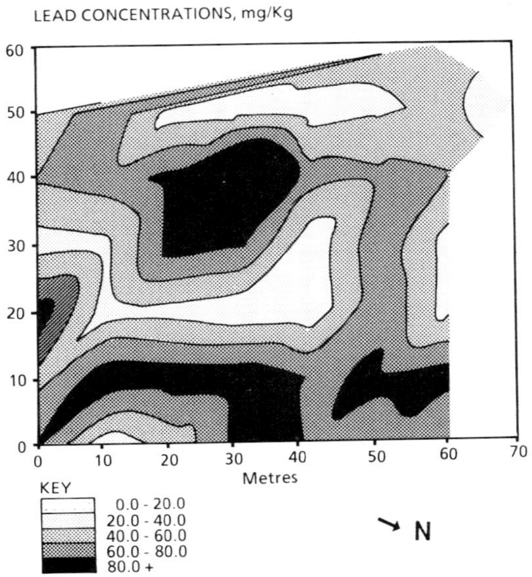

Figure 8 Distribution of lead concentrations around PP17.

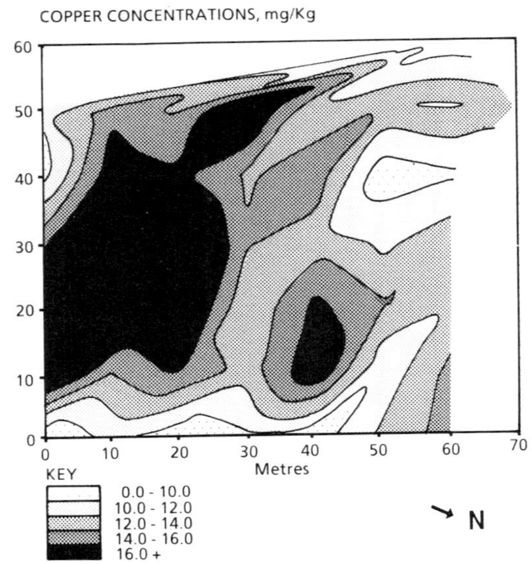

Figure 9 Distribution of copper concentrations around PP17.

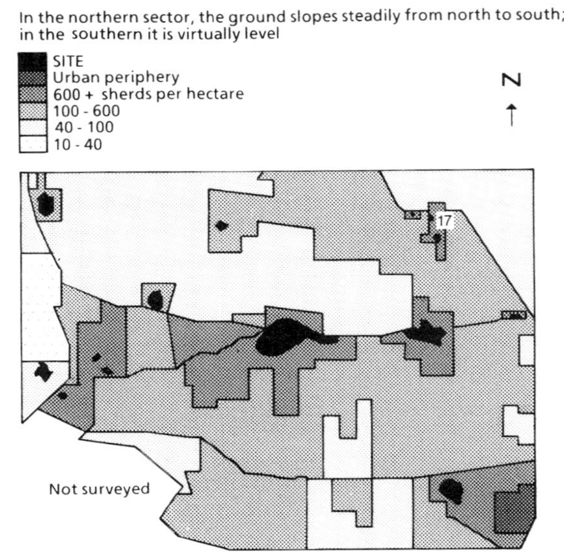

Figure 10 Pottery densities around PP17.

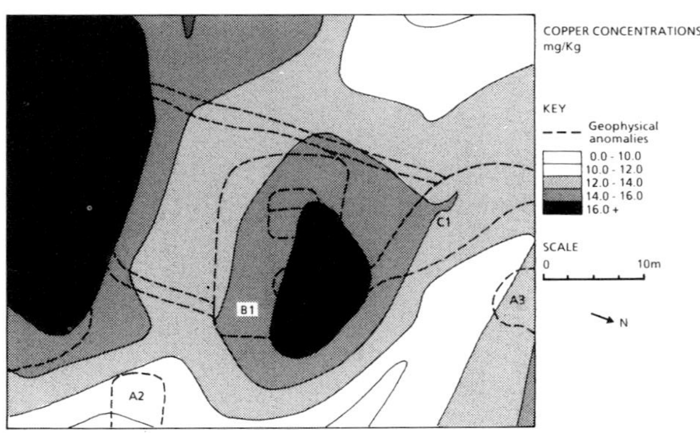

Figure 11 Copper concentrations and geophysical anomalies at PP17.

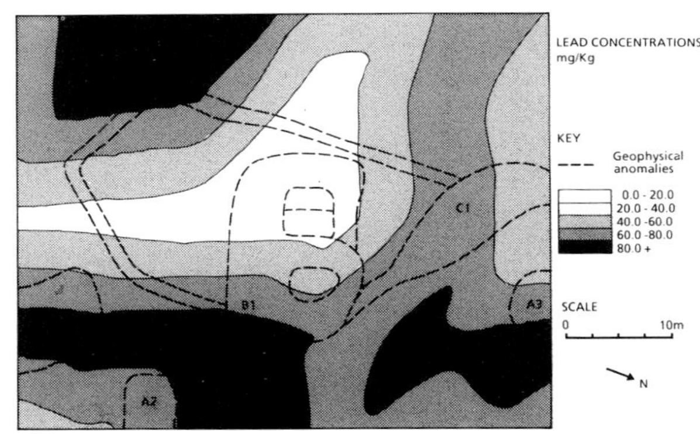

Figure 12 Lead concentrations and geophysical anomalies at PP17.

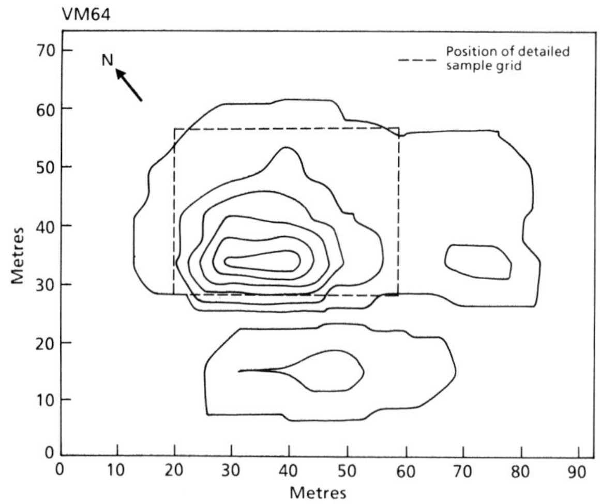

Figure 13 *Pottery concentrations around VM64.*

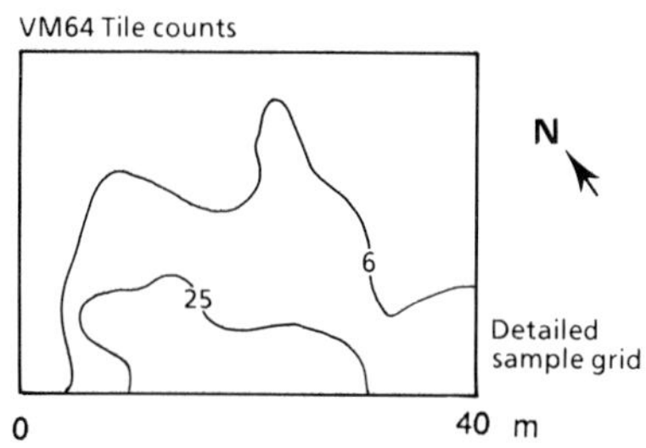

Figure 14 *Tile counts around VM64.*

Figure 15 *Interpretation of resistivity survey over the detailed sample grid at VM64.*

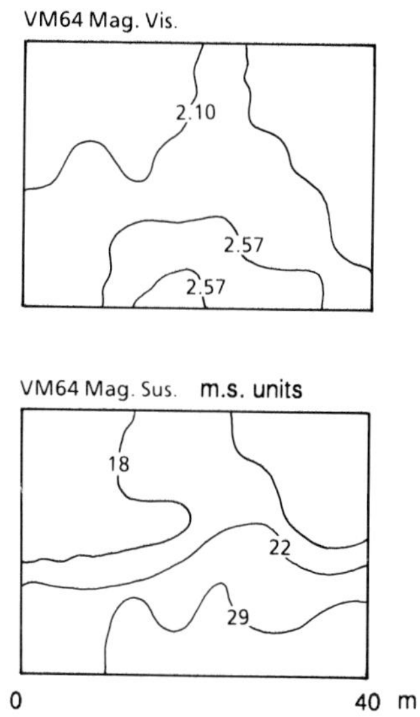

Figures 16 and 17 *Magnetic susceptibility and magnetic viscosity measurements over the detailed sample grid at VM64.*

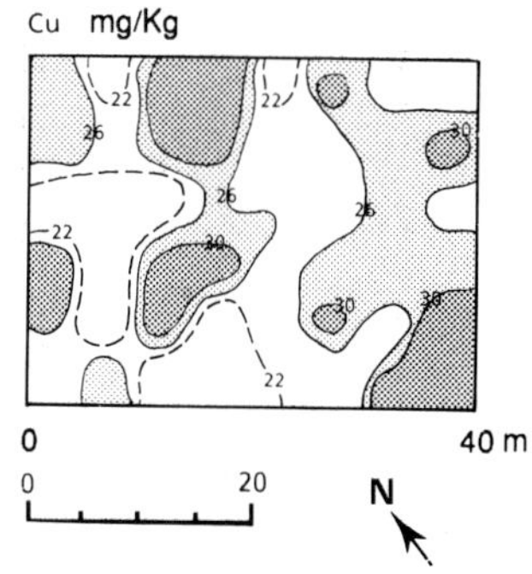

Figures 18 and 19 *Lead and copper concentrations over the detailed sample grid at VM64.*

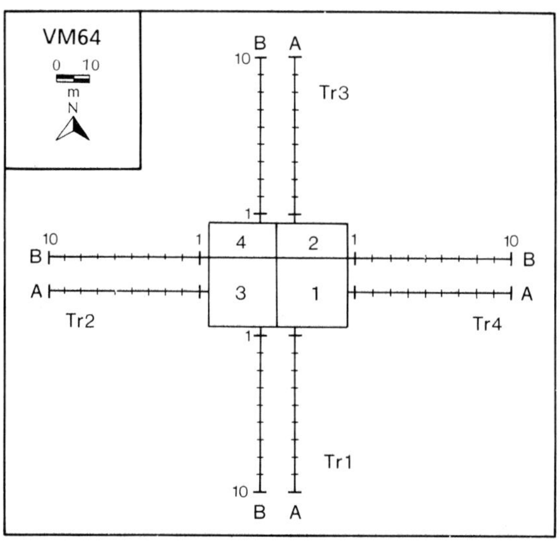

Figure 20 *Location of trace metal sampling transects around VM64.*

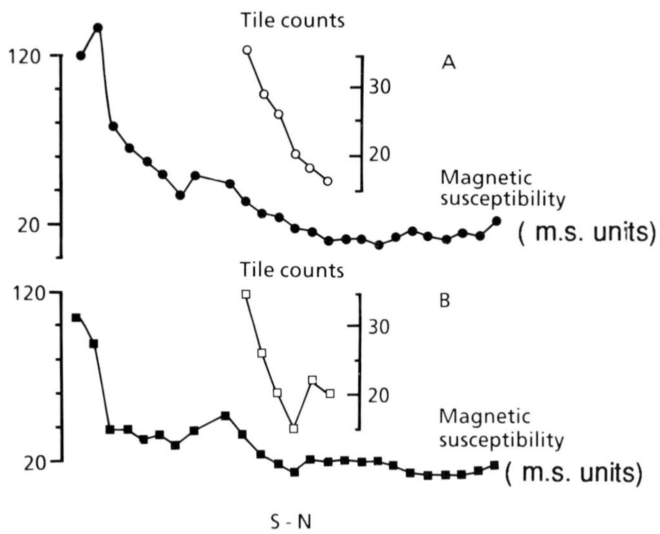

Figure 21 *Magnetic susceptibility measurements and tile counts along S-N transects, located as in Figure 20.*

Figure 23

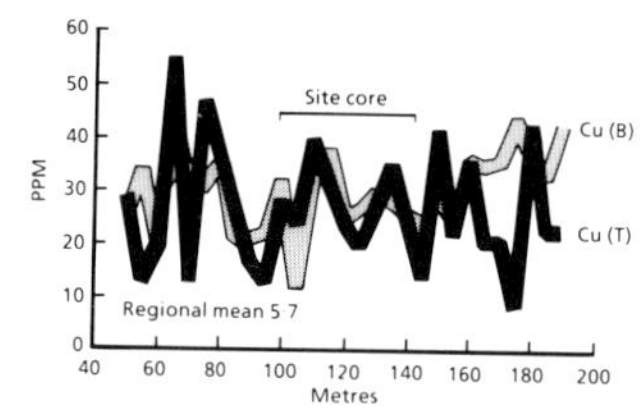

(a) Copper concentrations along W-E transects located as in Figure 20.

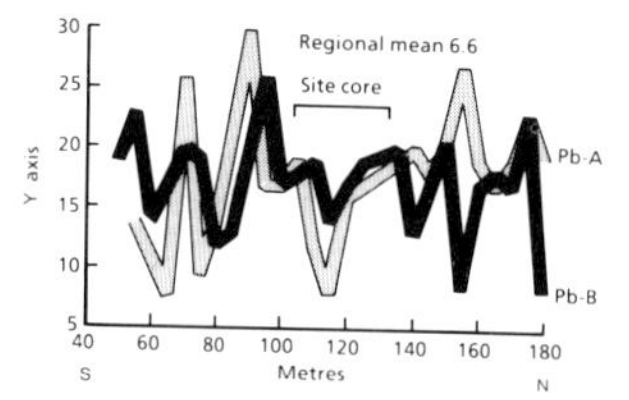

(b) Lead concentrations along S-N transects located as in Figure 20.

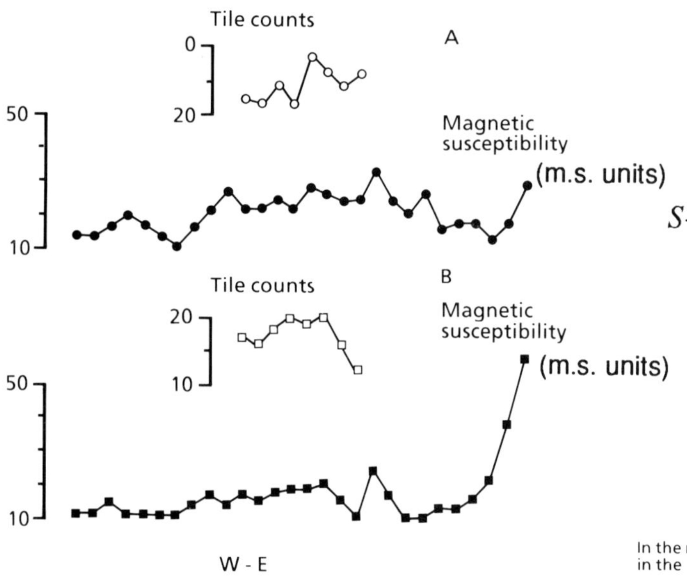

Figure 22 *Magnetic susceptibility measurements and tile counts along W-E transects, located as in Figure 20.*

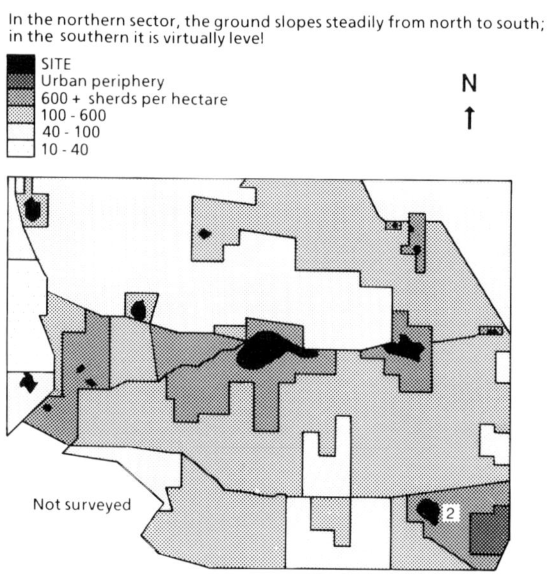

Figure 25 *Pottery densities aound TPW2.*

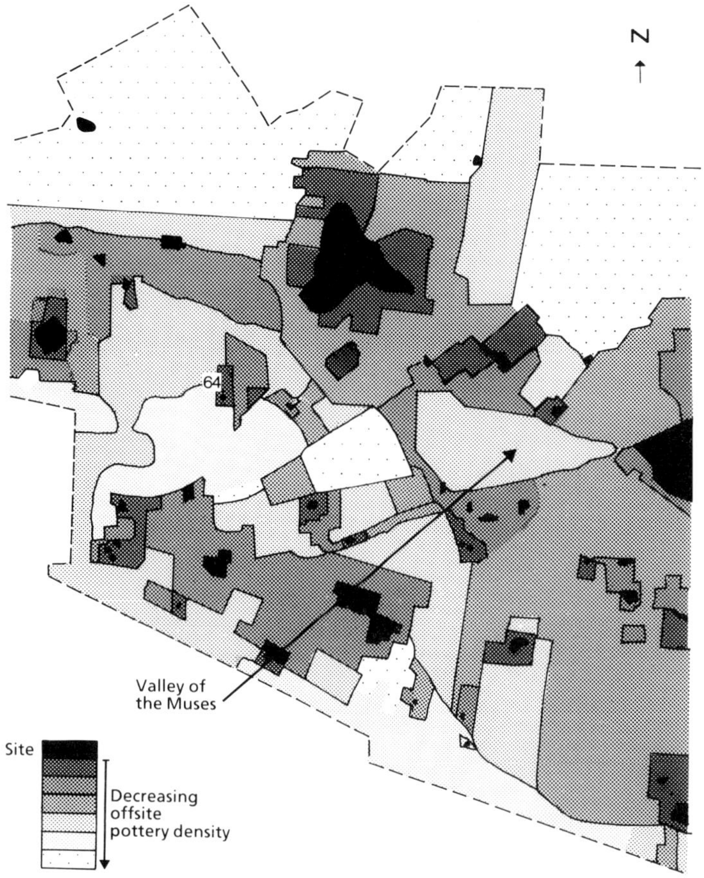

Figure 24 *Pottery densities around VM64.*

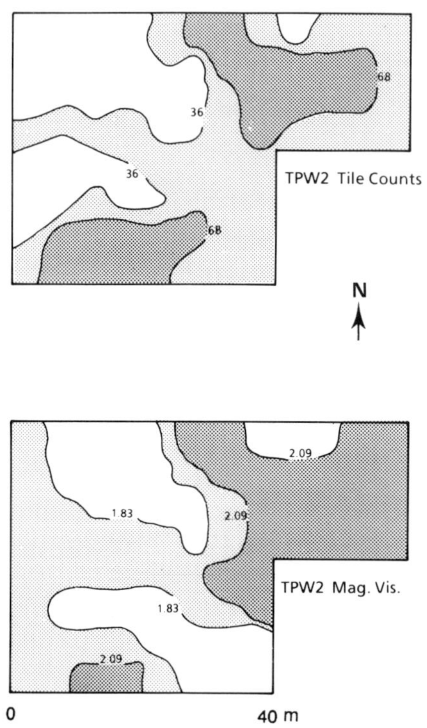

Figures 26 and 27 *Tile counts and magnetic viscosity measurements at TPW2.*

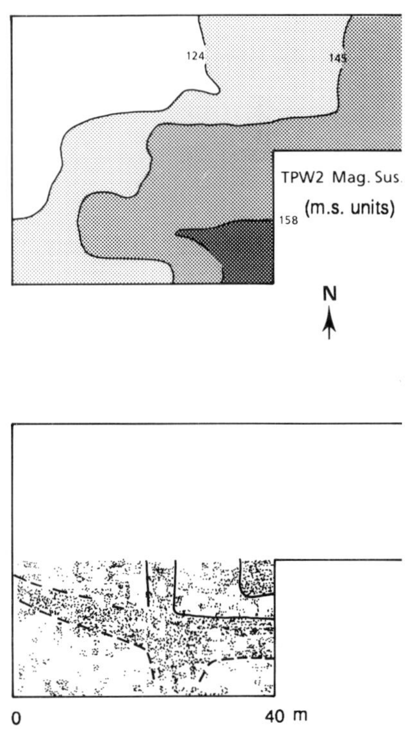

Figures 28 and 29 *Magnetic susceptibilit measurements and resistivity surve interpretation at TPW2*

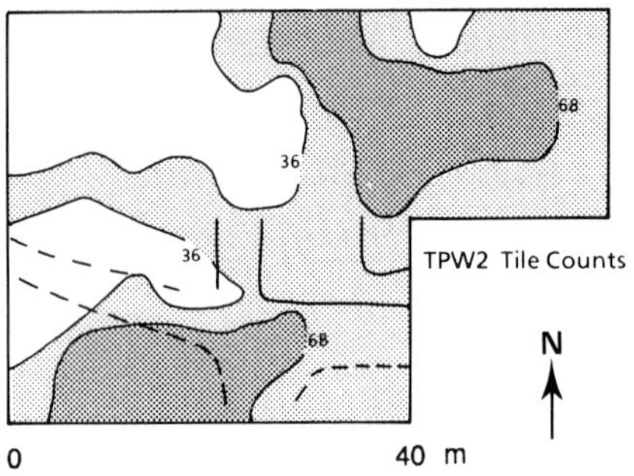

Figure 30 *Tile counts and resistivity survey features at TPW2.*

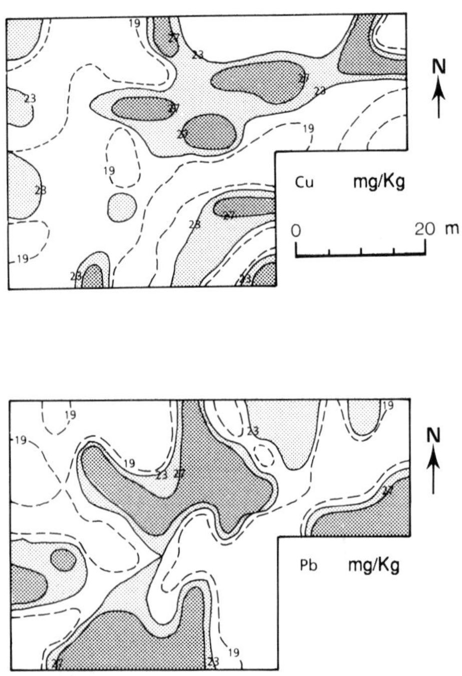

Figures 31 and 32 *Copper and lead concentrations at TPW2.*

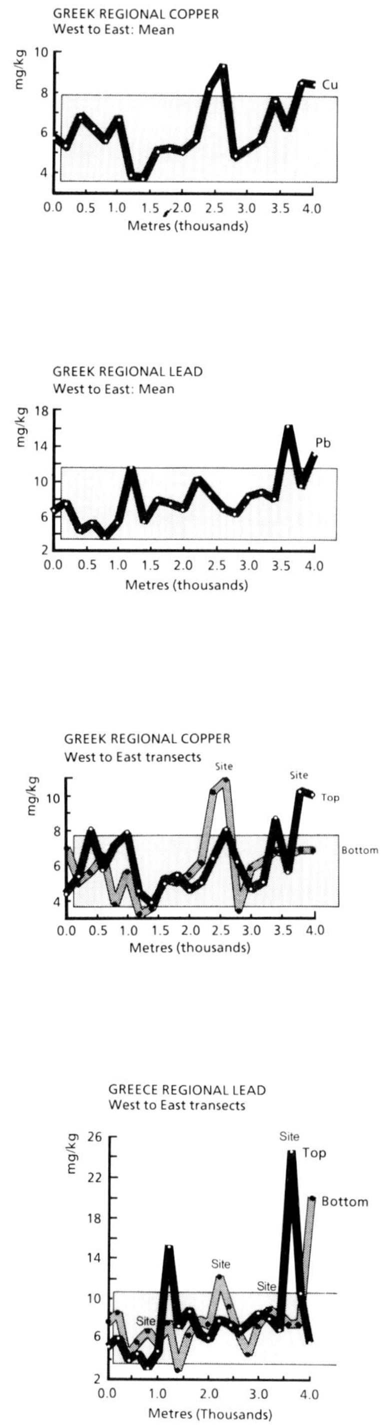

Figures 33 to 36 *Copper and lead concentrations (measured and mean) along West to East transects located as in Figure 2.*

Tofts Ness: Exploration and Interpretation in a Prehistoric Landscape

S.J. Dockrill* and J.A. Gater**

**Department of Archaeological Sciences, University of Bradford, Bradford BD7 1DP.*
***Geophysical Surveys, The Old Sunday School, Kipping Lane, Thornton, Bradford, BD13 3EL.*

Abstract

The extensive use of geophysical techniques on the Tofts Ness peninsula, Sanday, Orkney has greatly enhanced the interpretation of both the landscape and the paleo-economy of the main excavated site, providing an important development in the role of geophysical survey from that of the traditional one of site identification. Magnetic susceptibility measurements have been instrumental in suggesting, a) the source of Neolithic fuel, and b) evidence of early soil management. Off-site geophysical survey has revealed further evidence of land management identifying extensive deposits of anthropogenic soils. Geophysical techniques are seen as a fundamental part of the project and their integration with other techniques both on and off site is seen as an important aid to the interpretation of the surrounding paleo-landscape.

Introduction

Tofts Ness is a machair covered peninsula on the northern arm of the isle of Sanday, Orkney (Fig. 37). The area is bounded by a dune system to the southeast and an inland loch to the south. Much of the area is below the 5 metre contour. An extensive archaeological landscape survives composed of mounds, enclosures, banks and cairns. The main concentration of sites was surveyed in 1981 by Jack Stevenson of the RCHAMS, and has since been granted scheduled monument status by Historic Buildings and Monuments branch of the Scottish Development Department. The outlying monuments are under threat from clearance, ploughing (for re-seeding) and mole ploughing.

The aims of the Tofts Ness Project include the evaluation of the archaeological potential of the Tofts Ness peninsula, and the assessment of the level of contemporary threat to the archaeology of the area. The detailed excavation of one threatened site was included as part of the project's goals, whilst geophysical techniques were considered integral to research work both off, and on, site. (1)

The excavation and its geophysical context

The excavation strategy applied on the threatened site selected for detailed investigation included:

1) The investigation of a settlement mound in order to gauge the nature and extent of surviving deposits, and the relationship between the site and the surrounding landscape.

2) The area excavation of a circular stone structure within the mound, which had been partially exposed as the result of the burrowing activities of rabbits.

Full magnetic and resistivity surveys were carried out in advance of this work.

Excavation proved the settlement mound to be multi-period with the earliest occupation on the site dating to the Neolithic c.3000BC, and with later evidence of occupation in the Early Bronze Age, Late Bronze Age and the Early Iron Age.

The area excavation of the circular stone structure revealed a complex dating to the Early Iron Age. The relationship of the excavation and geophysical surveys carried out over the site can be seen by comparing the magnetic data in Figure 38, which is displayed as a simple contour plan, with the excavation plan in Figure 39.

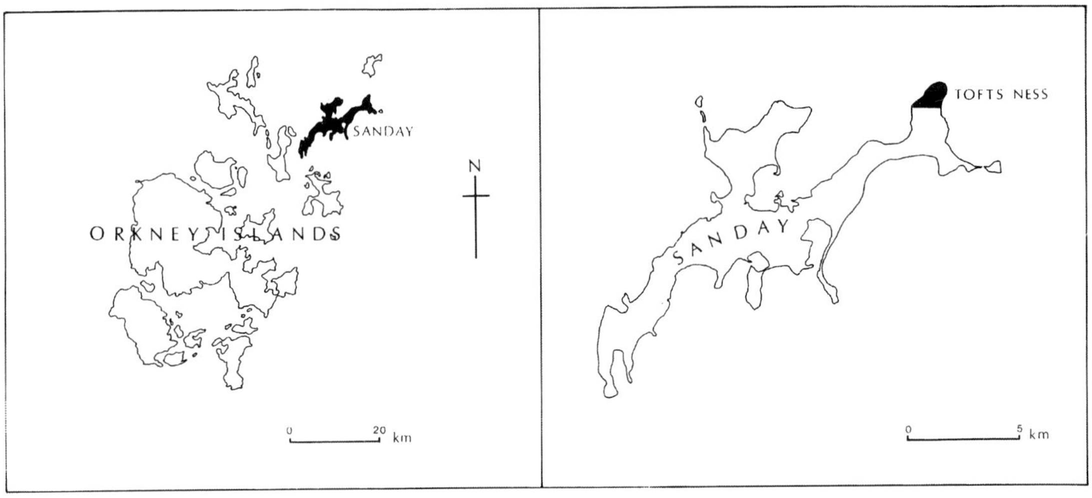

Figure 37 *Location of Tofts Ness, Sanday, Orkney.*

The positive anomalies of the magnetic survey of the round house complex are seen to reflect midden dumps outside of the building, and midden contained within wall cores. A discrete central anomaly reflects the central hearth. The magnetic anomaly to the east of the structure is seen to correspond exactly with a midden dump located through excavation.

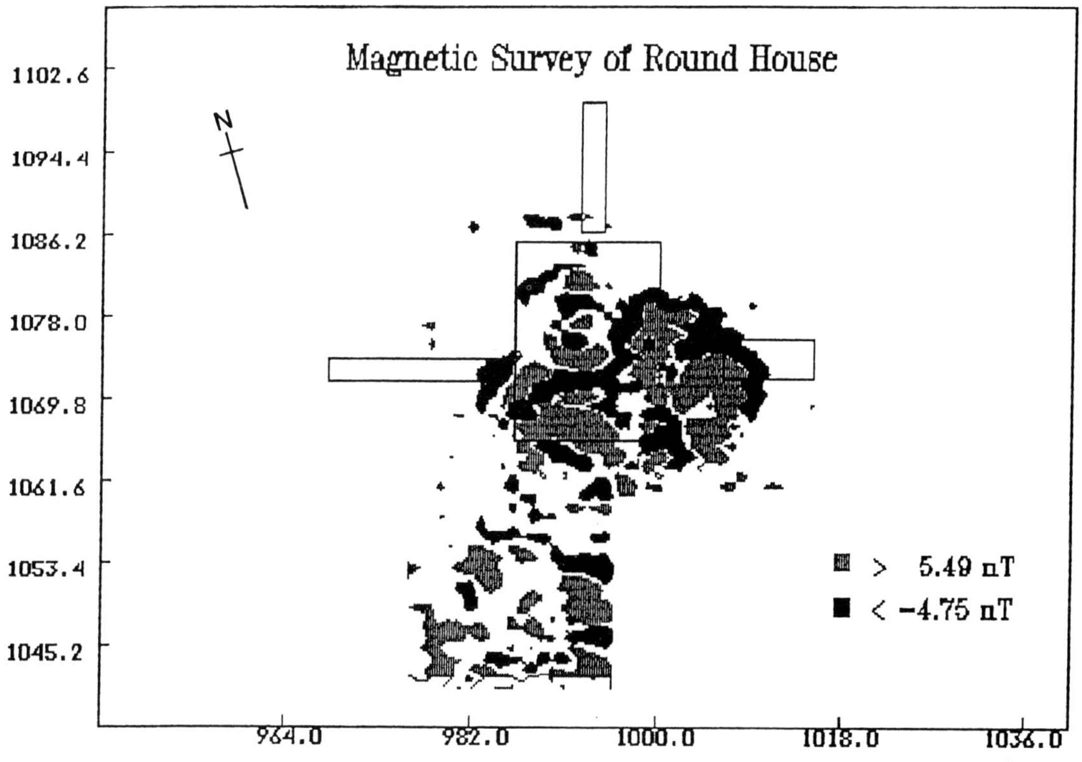

Figure 38 Magnetic data from the area of the round-house excavation. The trench plan is superimposed upon the one positive and one negative contour.

A column of magnetic susceptibility samples taken through the stratigraphic sequence of the midden dump to the east of the structure reveals a number of interesting points. Initially, it can be said that magnetic susceptibility samples revealed values ranging from 352x10-6 to 884x10-6 emu/gm. A buried sand-based soil with evidence of ard cultivation sealed by the midden tips provided a very low response of magnetic susceptibility of 4x10-6 emu/gm, indicating little application of manure. However, an earlier buried soil, cut by ard marks from the sand-based agricultural soil has an enhanced magnetic susceptibility of 136x10-6 emu/gm, indicating an anthropogenic origin for the soil. This suggests the application of material to the soil, indicating a high degree of land management. It would also suggest permanent, rather than seasonal, exploitation of the paleo-landscape in the early phase of settlement. At present these soils are being further studied using soil micro-morphological techniques.

The standard application of magnetic susceptibility analysis on soil samples from excavation contexts has also provided further insights into the nature of soils on the site. The distinctive reddish soil matrix of early bronze age midden spreads proved to have extremely high magnetic susceptibility. The high magnetic susceptibility values are due to the high iron content of the soil.

One interpretation of this phenomenon is that the matrix was formed by the burning of turf for fuel. A similar type of peaty turf deposit forming today on Tofts Ness was identified by Dr Susan Limbrey; a ferruginous peaty turf, rich in iron but with a low magnetic susceptibility. However, with the burning of a sample of this deposit, the iron was changed to a magnetic form comparable to the archaeological midden matrix.

***Figure 39** Excavated plan showing the final phase of the round-house complex.*

The use of magnetic susceptibility, as illustrated above, is seen as an important tool to be integrated with conventional environmental methods in aiding the reconstruction of human activities.

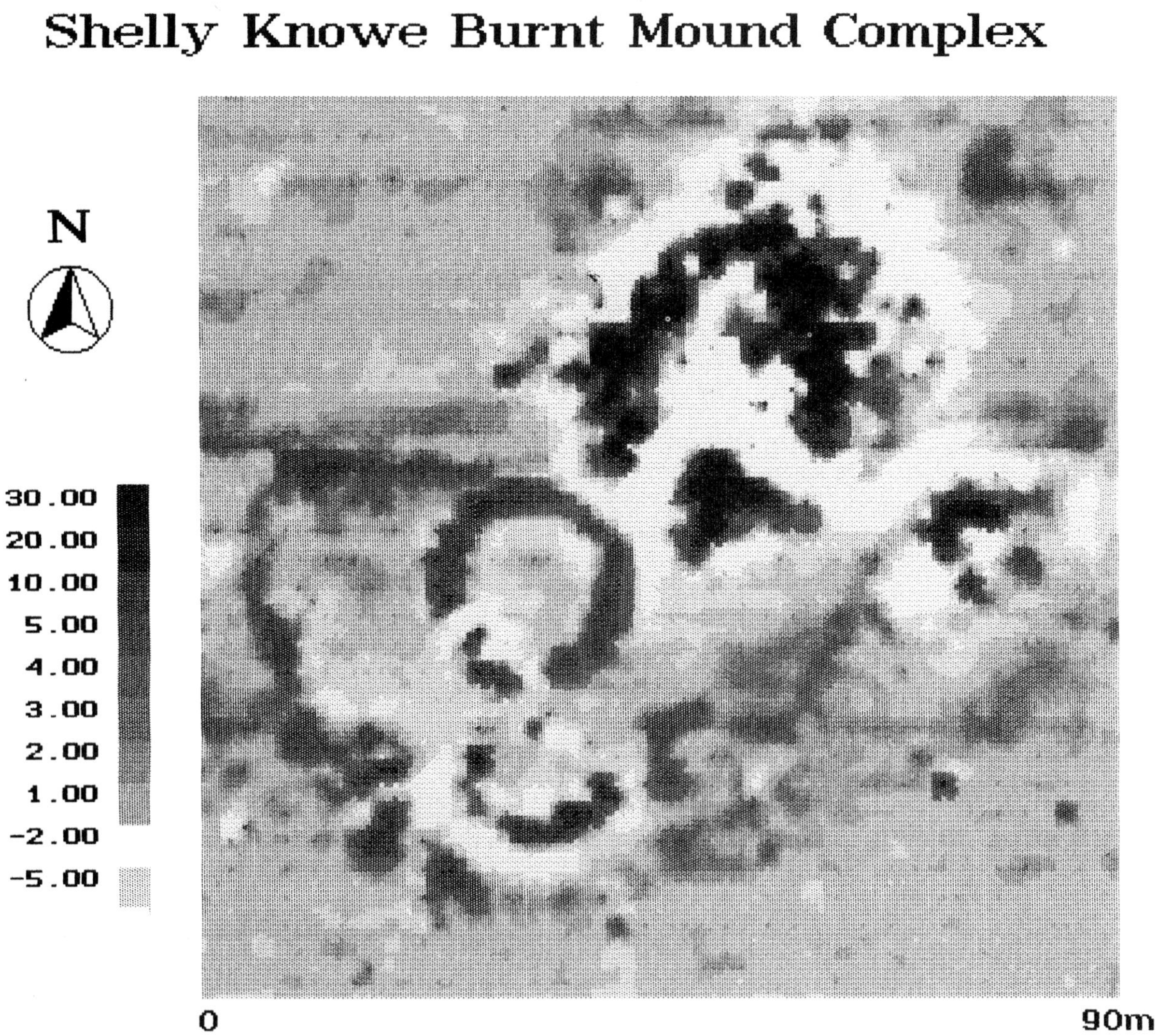

Figure 40 *Grey-scale representation of magnetic data from the burnt mound complex, indicating the crescent-shaped mound to the north-east and two circular structures to the south.*

Geophysical Survey in the Landscape

Within environments like that of Tofts Ness, earthwork morphologies often appear ubiquitous and defy attempts at qualification or characterisation. In some instances geophysical survey can be used, with great effect, to define site types or functions. A clear example of the use of geophysics in such a context can be illustrated through the work on the Shelly Knowe group of earthworks (Figure 40).

Here a burnt mound complex is clearly identified in the top right of the magnetic survey and is seen in contrast with midden deposits forming wallcore material in two circular structures to the south. The use of geophysical survey in this way has enabled the division of a number of amorphous grass-covered mounds throughout the peninsula into three categories: Settlement sites, burnt mounds and burial sites.

However, geophysical survey has the potential of application beyond the confines of the site. In the landscape surrounding the Shelly Knowe complex, geophysical survey has mapped an area of magnetic enhancement around the site, which has been cut by a later rig field system (Figure 41). The strong anomalies forming the boundaries of this area of enhancement are caused by an edge effect. Coring and excavation have revealed that the area of enhancement corresponds to an anthropogenic soil sandwiched between layers of windblown sand. The diagonal striations representing the later rig system are clearly shown again as edge effects due to the disturbance of the underlying 'uniform' archaeological soil with the cutting of the rigs.

The Burnt Mound Complex And Surrounding Landscape At Shelly Knowe

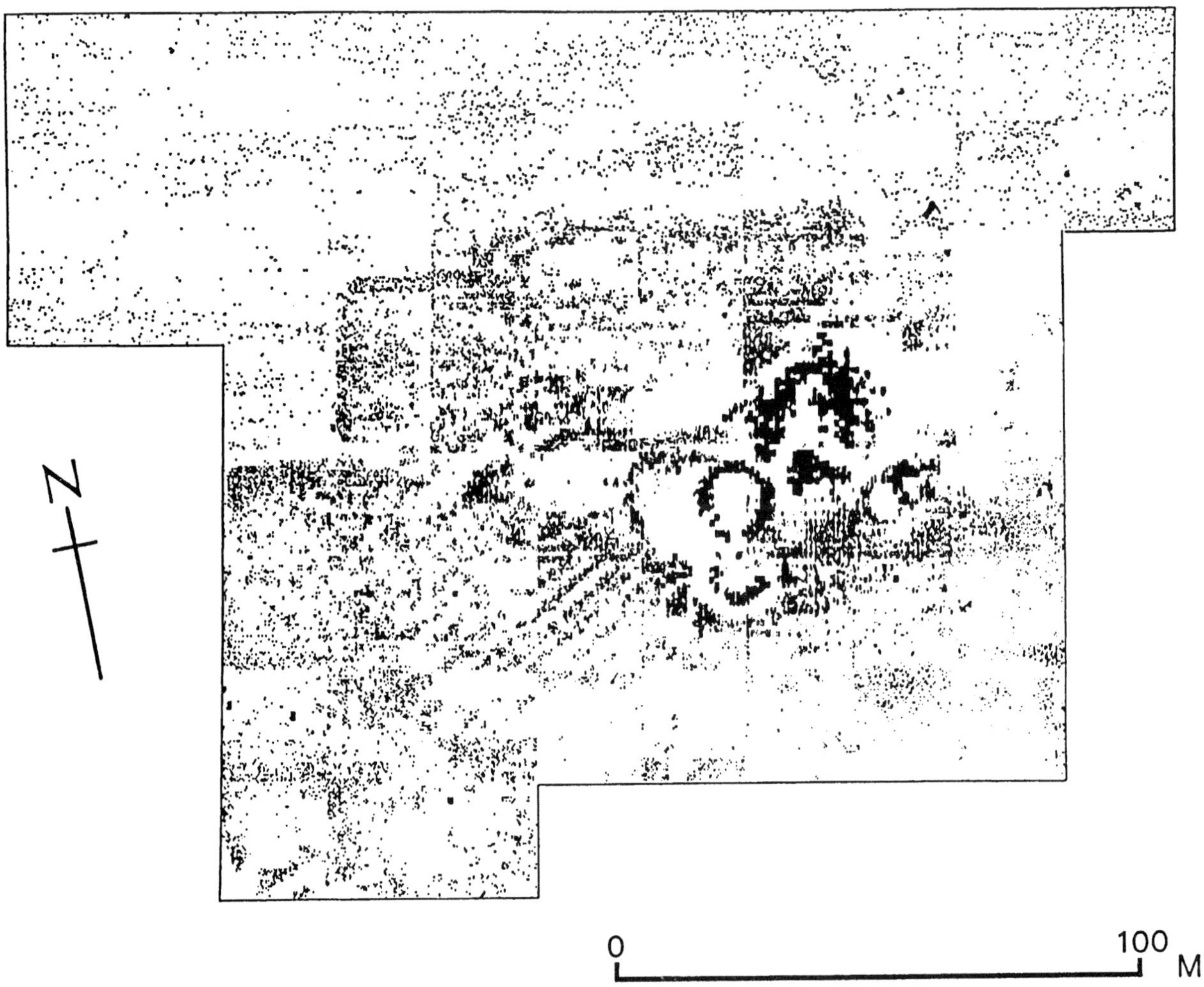

Figure 41 *Dot-density distribution plot of positive anomalies within the landscape surrounding the burnt mound complex.*

The presence of anthropogenic soils surrounding this site is not an isolated occurrence, it is repeated across the peninsula and it has been observed in the magnetic survey data and verified by coring and excavation. This data suggests areas of intensive arable activity with evidence of prehistoric soil management confined around areas of settlement. The completion of the examination of botanical, faunal and molluscan remains together with the micro-morphological study of the buried soils will allow, in the near future, a full evaluation of changes in economy and land use within the prehistoric period at Tofts Ness.

In conclusion, geophsical survey is seen as an important element within the Tofts Ness project, and has been successfully used beyond the 'traditional' role of site location. The integration of this data with that derived from conventional excavation and survey methods has allowed for a wider understanding of the archaeological landscape, together with its past exploitation by man.

(1) Resistivity survey work commenced in 1984 using a Geoscan RM4 Resistivity Meter, funded by the Society of Antiquaries of Scotland. The excavation project was directed by S.J. Dockrill for the Historic Buildings and Monuments branch of the Scottish Development Department and supported by the Department of Archaeological Sciences of the University of Bradford. The geophysical work was carried out by J.A. Gater using the following instrumentation: A Geoscan RM4 linked to a DL10 data-logger for resistivity survey, and several different fluxgate gradiometers for the magnetic survey: A Plessey 1.0m instrument, a Philpot AM01 linked to a DL10 (in 1986), and a Geoscan FM36 (in 1987).

Magnetic susceptibility samples have been taken throughout the course of the project, with measurements being recorded using a University of Bradford laboratory-based system. The authors would like to acknowledge the contribution of the late Andrew Jagger in this area.

Acknowledgements

The authors would like to thank the following for their help and use of software: Dr A.R. Walker (Geoscan), R. Trainer (Strata Software) and J. Haigh (University of Bradford). In addition, special thanks are due to Vince Gaffney for commenting upon the final draft of this paper.

Field Magnetic Susceptibility Measurement for Prospection and Excavation

Adrian Challands

The Old School House, Glinton Road, Helpston, Peterborough, PE6 7DG

Abstract

The potential of field magnetic susceptibility for both survey and excavation is examined. Magnetometry and magnetic susceptibility are compared. Chemical and physical principles are briefly described. Practical methods and pitfalls of surface survey are addressed. A surface survey undertaken in Orkney has been used as a practical example, and a small area excavated shows that higher readings on the surface are reflected in the underlying features. The value of magnetic susceptibility techniques in site interpretation is described using two different sites - one a barrow in Cambridgeshire sited on gravel, the other located within a sandstone Neolithic house on Orkney. The vexed question of magnetic susceptibility and phosphate intensity is briefly referred to using the Orkney example.

Introduction

Amongst the prospection techniques commonly used for archaeological reconnaissance, field magnetic susceptibility measurement is one of the lesser used methods. This short paper is written to describe some of the methods, pitfalls and potential for practical archaeology of field magnetic susceptibility techniques.

Dr. A.J. Clark's paper "The testimony of the topsoil" (Clark 1983) puts a strong case for a detailed geophysical/geochemical examination of the topsoil to enhance and evaluate aerial photographic evidence. Magnetic susceptibility is one of the techniques which Dr.Clark discusses-although the examples which he quotes are based mainly on laboratory measurements.

Laboratory measurements of the magnetic susceptibility of soils and sediments have been carried out over a number of years-particularly as part of the analysis of, for example, lake sediments. Laboratory testing of magnetic susceptibility, whilst quantitative and accurate, is a relatively slow process a soil/sediment sample has to be air dried, crushed and sieved before measurement in a laboratory coil. For archaeological reconnaissance purposes, field measurements of magnetic susceptibility whilst only relative and not quantitative, may, more quickly, provide an equally acceptable evaluation-as a large number of readings can be taken rapidly over a survey area.

What are the advantages of magnetic susceptibility survey when compared against magnetometry? The main advantage is that magnetic susceptibility measuring equipment generates its own magnetic field which is used to measure the magnitude of the magnetically enhanced minerals within a soil. Since human occupation can cause magnetic enhancement of soils, it follows that field magnetic susceptibility is a useful technique for locating the extent of, say, earlier prehistoric settlement areas which have very small or no subsoil cut features. Magnetometry, using either a proton or fluxgate magnetometer, will normally only detect local changes in the Earth's magnetic field-caused by silt/enhanced soil filled features cut into the subsoil or the usually strongly magnetic fired clay of kilns and hearths.

Basic principles

It is beyond the scope of this paper to discuss the chemical changes and physical principles involved in magnetic susceptibility techniques-detailed information is contained in the published works of Thompson and Oldfield (1986), Scollar and Graham (1972) and Tite and Mullins (1971). The basic principles required in order to appreciate the potential of field magnetic susceptibility measurement for archaeological purposes, are given below.

Magnetic susceptibility may be defined as the ratio of the induced magnetic field in a material to the applied magnetic field. Iron oxides are contained in most of the Earth's soils, usually in the form of weakly magnetic haematite (αFe_2O_3). When haematite (αFe_2O_3) is subjected to heating (generally burning) conversion occurs to strongly ferrimagnetic magnetite (Fe_3O_4) which when re-oxidised, upon cooling, converts to ferrimagnetic maghaemite (γFe_2O_3) (Le Borgne 1955, 1960).

Generally, the magnetic susceptibility of topsoil is higher than the subsoil from which it is derived (Le Borgne 1955). But if the enhanced soils of pits and ditches are ploughed up from relatively flat occupation surfaces, increased localised magnetic susceptibility may be detected in the topsoil lying over the buried archaeological features.

It should be pointed out that "the amount of iron oxide available in the soil for conversion to maghaemite determines the susceptibility acquired by the layer of soil which is heated" (Tite and Mullins 1971). For example, field magnetic susceptibility measurements may be reduced over chalk subsoils (Aitken 1974, 222-3).

Practical field magnetic susceptibility

i) Instrumentation

The magnetic susceptibility measuring instrument commonly used for archaeological purposes is the Bartington MS1 magnetic susceptibility meter*, together with a 180 millimetre diameter search loop or field coil. A ground penetration of up to 120 millimetres may be achieved below and around the annular field coil. The Bartington MS1 meter can be connected to a laboratory coil sensor, the MS1B, in order to measure the absolute magnetic susceptibility of a sample of known mass and volume.

ii) Surface prospection

When using field magnetic susceptibility measurement for archaeological prospection a number of factors must be taken into account.For example,it must be constantly borne in mind that the magnetic susceptibility value obtained from a field coil is not an absolute measure of the magnetic susceptibility, because the mass and volume of the soil at the point of measurement cannot be determined in the field. Therefore it is more correct to refer to field values as MS UNITS. Field magnetic susceptibility readings are only relative,either one to another ,or each to a series of background readings. For practical prospection purposes, field magnetic susceptibility readings are usually taken on a grid pattern-mostly at 1 or 0.5 metre increments. This grid pattern enables the data generated to be easily processed statistically and displayed as a dot density or symbol diagram using a personal computer. It is also good practice to take soil samples at, say, every 50 field readings (depending on the size of the survey area and soil changes), for absolute measurement of magnetic susceptibility in a laboratory coil in order to provide both a check on the field data and to assist the later interpretation of the survey data.

The status of the field surface should also be taken into account when undertaking a field magnetic susceptibility survey. A flattish surface is required to obtain an even contact between the soil and the field coil. The ideal surface is after the land has been ploughed, harrowed and weathered. Freshly ploughed land can pose severe coil contact problems since the vertical distance from the top to bottom of the furrow usually exceeds the 120 millimetres penetration, and the coil is too large to manoeuvre within the furrow. A small coil, 15 millimetres diameter is available for use in these conditions, but on a large survey it can be rather slow to use, and the readings are less consistent.

It is as well to be aware of the effects of soil compaction when undertaking a surface survey over arable land. In the laboratory the absolute magnetic susceptibility value is measured and calculated taking into account the mass and volume of the sample.

....

* *The Bartington MS1 has now been superseded and the improved MS2 range also has data logging facilities. For further details contact Bartington Instruments Ltd., Spendlove Centre, Enstone Road, Charlbury, Oxford. OX7 3PQ.*

It follows that in the field, localised compaction, perhaps by tractor tyres, will result in a greater mass for a given volume, thereby giving a higher reading over the compacted area. Where localised compaction occurs on a survey location, it is prudent to slightly displace the field coil to an uncompacted location.

In order to demonstrate the potential of field magnetic susceptibility surface survey, a survey carried out in May, 1987 at the Barnhouse neolithic settlement site, Stenness Orkney for the Scottish Development Department, will be used as an example.

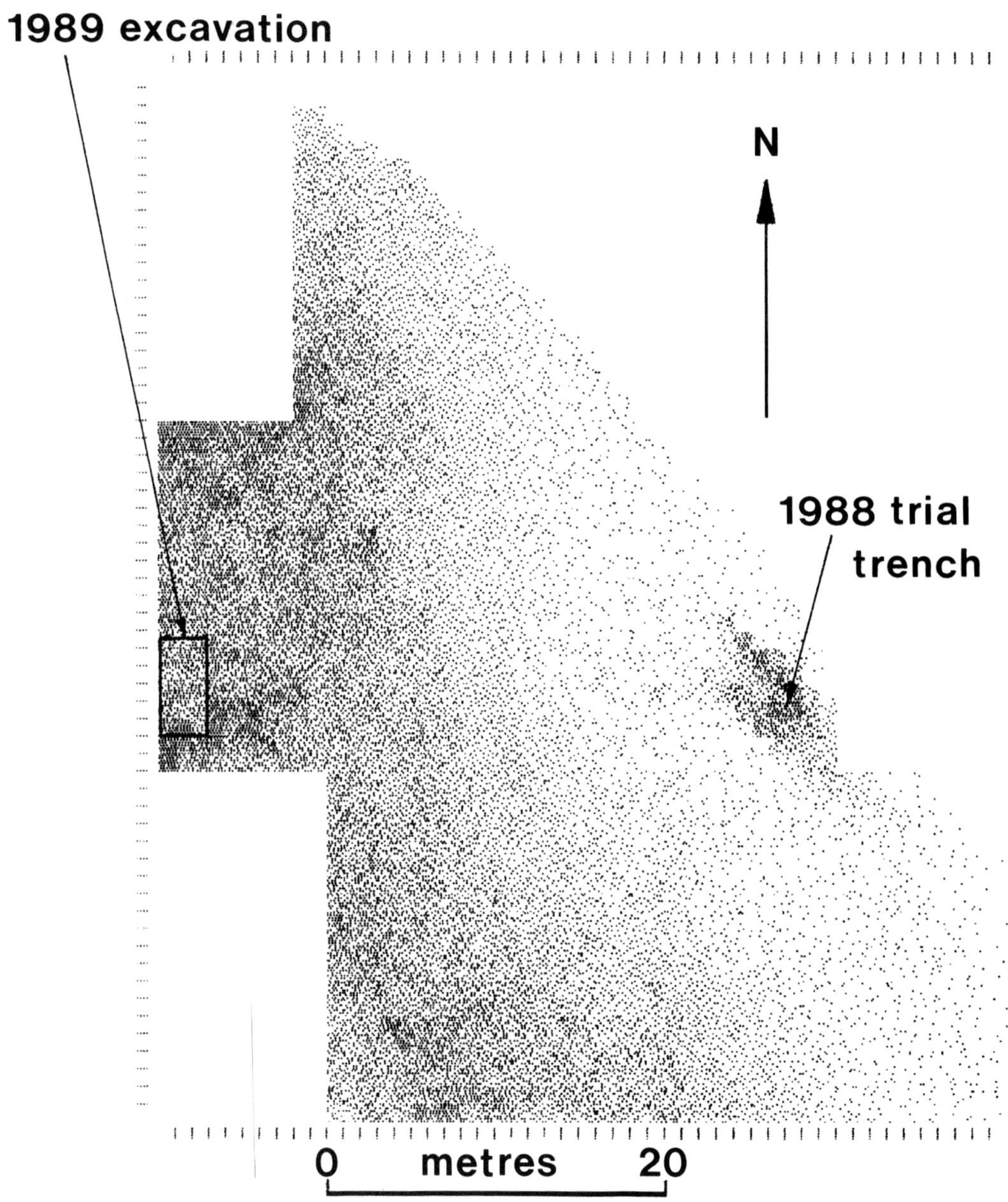

Figure 42 Dot density diagram displaying surface readings of magnetic susceptibility intensity at Barnhouse, Orkney.

At Barnhouse, magnetic susceptibility was employed to define the eastern extent, as far as the loch edge, of the unexcavated settlement area. The existing site grid was projected to form the survey grid, although spoil heaps lay between the area excavated in 1986 and the survey location. Consequently it was not possible to directly link the survey area and the 1986 excavated site. The field surface was at its optimum condition, ie. harrowed and weathered, for field magnetic susceptibility measurement. Field measurements were taken at 1 metre increments and logged in 20 metre squares in order to process the data on site using an Epson HX20 portable computer and a programme developed by Dr.M.Kelly. The dot density diagram Figure 42 is a mosaic of the 20 metre squares merged and displayed at a uniform range by means of a programme developed by Robert Randall for the IBM PC. Figure 42 shows increased magnetic susceptibility readings up to 20 metres away from the excavated settlement site and a gradual dilution of intensity towards the loch edge. A discrete area of enhancement may be seen on the loch edge which was sectioned in 1988 and proved to have human occupation layers so plough damaged that further work was not justified. In 1989 a small area was excavated (Figure 42) within an area of high surface magnetic susceptibility near the previously excavated area, which proved that the surface magnetic susceptibility survey accurately reflected the stratified settlement deposits. After the removal of the topsoil, magnetic susceptibility readings were taken, using a field coil, at 0.5 metre intervals to form a grid of magnetic susceptibility units. Figure 43 shows the readings superimposed over a simplified site plan. Magnetic susceptibility values of <45 units represent the natural subsoil and >100 units represent varying densities of neolithic occupation deposits.

iii) On site use of field magnetic susceptibility

It is well known that field magnetic susceptibility measurement using a probe type sensor can be useful in the interpretation of archaeological sections. But, the use of field magnetic susceptibility on an area basis on topsoil stripped archaeological sites is less well known.

Magnetic susceptibility used on archaeological sites on an area basis can materially assist in the rapid assessment of the features, particularly in identifying the locations of the most intensive human activity. The technique is especially useful on gravel sites where some periglacial features and archaeological features can look very similar. The magnetic susceptibility of periglacial features and natural gravel can be around 7 to 20 units, whilst archaeological features generally have a range of >50 units.

Field magnetic susceptibility measurements can also be very useful in the interpretation of complex archaeological structures. An example of magnetic susceptibility applied to a complex archaeological structure is the survey carried out in advance of archaeological excavation over a burial mound located on fen gravels. The burial mound lay on the line of a trunk road village bypass at Etton, Cambridgeshire. After removal of the topsoil the field magnetic susceptibility readings were taken at 0.5 metre increments. Figure 44 displays the data obtained as a dot density diagram which shows that the barrow is a composite construction. Although the final analysis of the excavation results has yet to be completed,including the interpretation of the geochemical and soil testing data, the magnetic susceptibility data (Figure 44) clearly shows that the barrow is either a two phase monument, with a primary small mound which was later enlarged, or a disc barrow. Grave 1 contained early bronze age pottery, although this overlay earlier beaker burials. The silty loam filled burials and the outline of the possible primary mound are represented by lower magnetic susceptibility values,although partially obscured by low values over the silt filled medieval ditch which cuts the barrow. Field magnetic susceptibility methods proved to be of value at Etton as some of the features, such as the possible primary mound, were visually difficult to discern on the stripped surface.

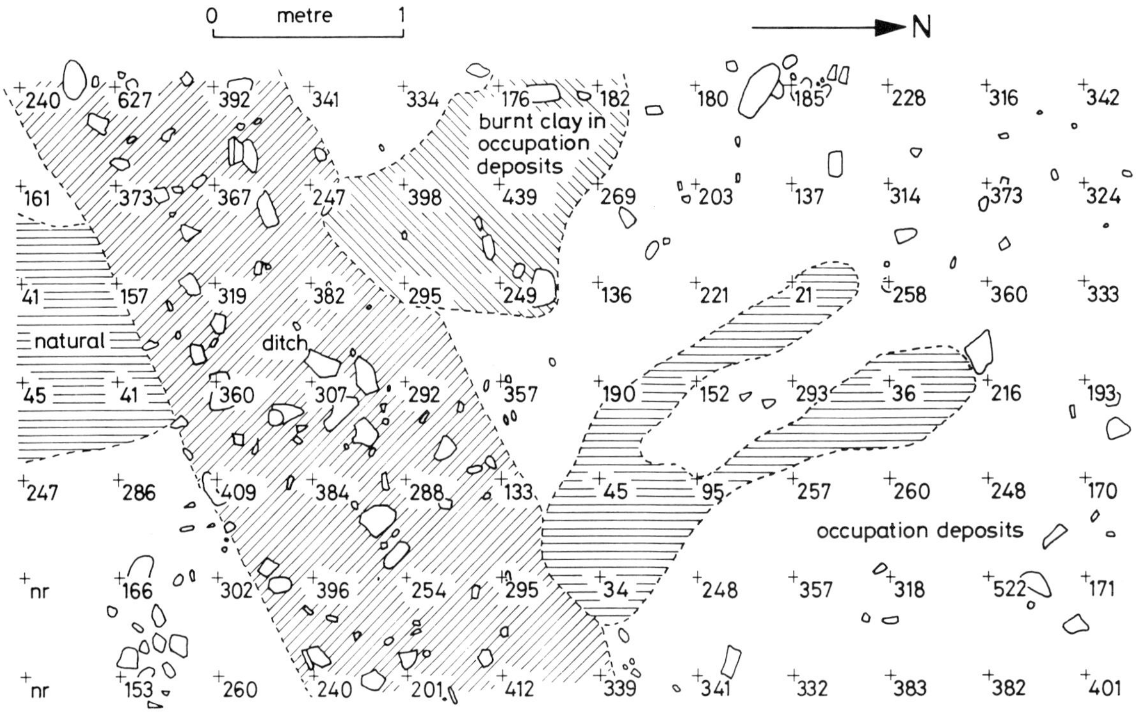

***Figure 43** Magnetic susceptibility units superimposed on simplified site plan of eastern trench at Barnhouse, Orkney.*

On site field magnetic susceptibility measurements were taken at 0.5 metre increments over all the excavated neolithic houses at Barnhouse, Orkney. At Barnhouse the neolithic houses are carefully constructed of Caithness flag. One house was larger and more carefully constructed than the others. It is this house which has been selected as another example of site magnetic susceptibility interpretation.

The interior plan of the house consists of a rectangular open area orientated north east-south west containing two, although not contemporary, square hearths lying central to the long axis of the open area. Around the open area six large recesses were set into the thickness of the Caithness flag walling: single recesses on the short axis and two recesses either side of the long axis. The Director, Colin Richards, puts forward the hypothesis that the house fulfilled a communal function, on the basis of the high quality construction and the cleanliness of the open "communal" area. Artifacts, burnt clay and organic debris were concentrated around the walls and in the recesses. Figure 45a supports this interpretation as high magnetic susceptibility readings cluster around the walls and recesses with low magnetic susceptibility readings in the open area, except, as is to be expected, in the area of the hearths where the readings are high.

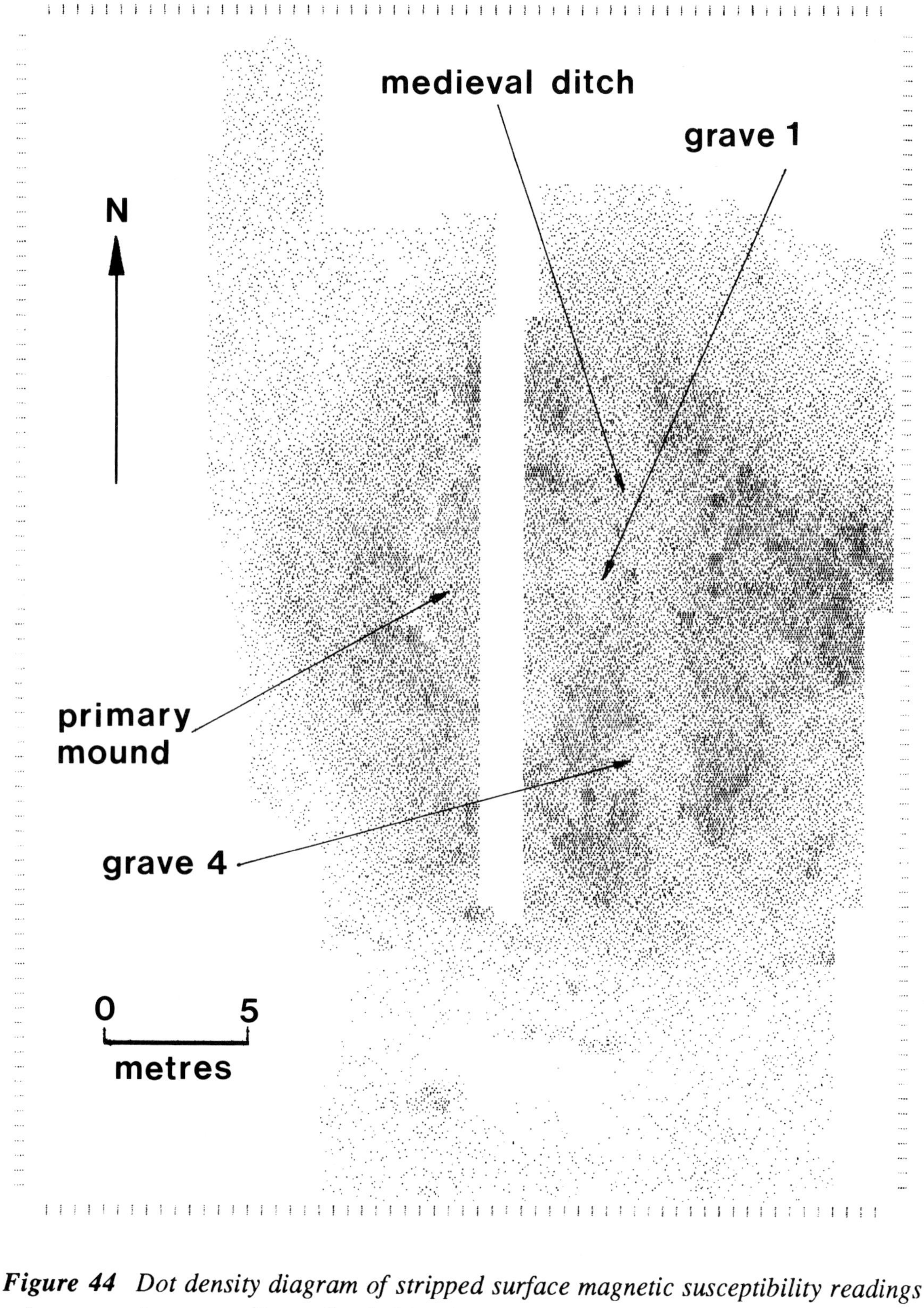

Figure 44 *Dot density diagram of stripped surface magnetic susceptibility readings taken over a barrow at Etton, Cambridgeshire.*

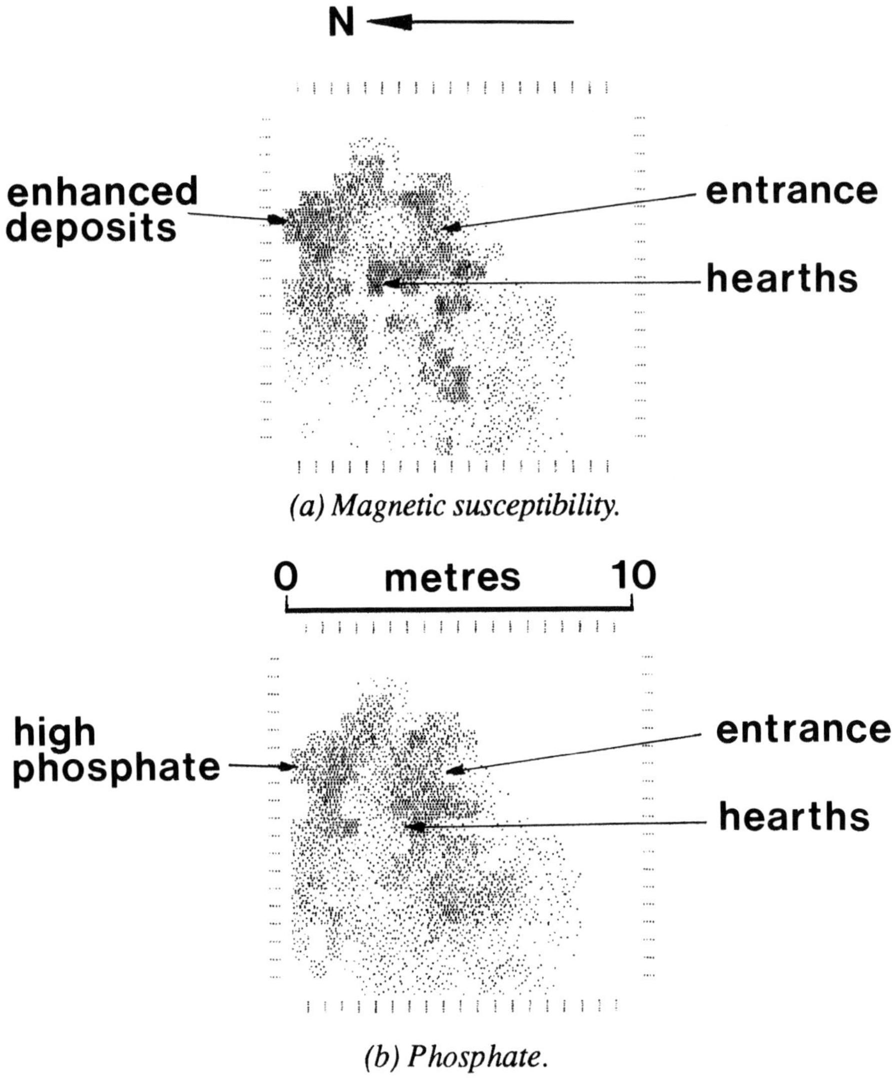

Figure 45 *Diagram showing (a) magnetic susceptibility readings and (b) phosphate intensities over an excavated neolithic house at Barnhouse, Orkney.*

It is interesting to compare the magnetic susceptibility dot density diagram Figure 45a with the phosphate intensity diagram Figure 45b where phosphate samples were taken at the same intervals as the magnetic susceptibility readings. Normally magnetic susceptibility readings and phosphate intensities are not compatible. But at Barnhouse it would seem that organic material (high phosphate) and burnt material (high magnetic susceptibility) were both cleared together to the sides of the building in the vicinity of the north eastern hearth.

In conclusion, field magnetic susceptibility has its place amongst the other geophysical methods used in archaeology. One advantage is that the magnetic susceptibility field coils penetration is to a known depth and is therefore unaffected by deeply buried magnetic features. Also, it is possible, by using a field coil, to survey at a rate of 400 readings in 15 minutes.

References

Aitken M., 1974, *Physics and Archaeology*, Oxford, Clarendon Press.

Clark A.J., 1983, 'The testimony of the topsoil', In G.S. Maxwell (ed.) *The impact of Aerial Reconnaissance on Archaeology*, C.B.A. Research Report No.**49**, London.

Le Borgne E., 1955, 'Susceptibilite magnetique anormale du sol superficiel', *Annales de Geophysique* **11**, 399-419.

Le Borgne E., 1960, 'Influence du feu sur les proprietes magnetiques du sol et sur celles du schiste et du granite', *Annales de Geophysique* **16**, 159-195

Scollar I.and Graham I., 1972, 'A method for the determination of the total magnetic moment of soil samples in a constant field', *Prospezioni Archeologiche* **7**, 85-92.

Thompson R. and Oldfield, F., 1986, *Environmental Magnetism*, Allen and Unwin, London.

Tite M.S. and Mullins C., 1971, 'Enhancement of the magnetic Susceptibility of soils on archaeological sites', *Archaeometry* **13**, 209-219.

Geoarchaeological Enhancement of River Valley Archaeology in North East England

Mark G. Macklin*, David G. Passmore*, David C. Cowley•
Tony C. Stevenson* and Colm F. O'Brien**

**Department of Geography, University of Newcastle upon Tyne.*
***Archaeological Practice, University of Newcastle upon Tyne.*
•RCAHMS, Edinburgh.

Abstract

Newcastle University's Geography Department and Archaeological Practice have been applying geoarchaeological methodologies to investigations of Holocene valley fill sediments in north east England both to elucidate Post-glacial alluvial histories and as a means of enhancing regional archaeological records. Archaeological linkages can often be made with alluvial and colluvial sedimentary sequences by virtue of the impact that anthropogenic activities have on rates of run-off, erosion and sediment production.

Preliminary results are presented from studies of a small upland catchment near Callaly Moor, Northumberland, and the lower Tyne valley near Newcastle upon Tyne. Contrasts in Holocene alluvial and colluvial sedimentary sequences are apparent both within and between catchments that appear to reflect differing anthropogenic, climatic and geomorphic histories of river basins.

Introduction

Archaeologists are becoming increasingly aware of the potential of Holocene alluvial and colluvial valley fills as sources of archaeological information. In part this is in response to the identification of settlement sites and artefacts buried beneath spreads of alluvium (e.g. Needham 1985) and colluvium (e.g. Bell 1981, 1982), and also the considerable opportunities such sedimentary sequences provide for palaeoenvironmental reconstructions that can be used to elucidate relationships between landscape history and human activity. Good examples of the 'geoarchaeological' approach are several recent studies of Holocene valley development in lowland (Brown 1983, Robinson and Lambrick 1984, Burrin and Scaife 1984) and upland (Harvey and Renwick 1987, Macklin and Lewin 1986, Richards *et al* 1987) catchments in England and Wales.

Since 1987, Newcastle University's Geography Department and Archaeological Practice (formerly the Archaeological Unit for North East England) have been applying geoarchaeological methodologies to investigations of river catchments in north east England to elucidate Post-glacial alluvial histories and to assess the archaeological potential of valley fill deposits. This paper outlines the conceptual framework we have adopted in our investigations and also presents some preliminary results from studies of upland and lowland alluvial valley floors in the region.

Archaeological Sediments in Holocene Valley Fills

Cultural linkages can often be made with off-site alluvial and colluvial sedimentary sequences by virtue of the demonstrable impact anthropogenic activities such as forest clearance, cultivation and mining have on rates of run-off, erosion and sediment production (Butzer 1982, Lewin and Macklin 1987, Thornes 1987). Geoarchaeological investigations of lowland river basins in southern England have identified widespread deforestation and arable cultivation during and after the second millenium bc as having a particularly significant impact on patterns and rates of valley floor alluviation (Brown and Barber 1985, Burrin and Scaife 1988, Shotton 1978). The consequences of these changes in land-use are often manifested in floodplain lithostratigraphy as episodes of accelerated deposition of fine-grained minerogenic sediments. These units contrast with underlying deposits of organic-rich silts, clays and gravels that were laid down earlier in the Holocene prior to large-scale anthropogenic disturbance. Upland catchments in Britain in this respect are less-well documented, although episodes of alluviation that appear to relate to human disturbances of catchment vegetation have been recorded in several smaller upland river systems (e.g. Harvey *et al* 1981, Macklin and Lewin 1986, Richards *et al* 1987).

It is apparent from these studies, however, that responses of river sediment systems to changes in land-use are conditioned by a complex interaction of factors that may be subject to considerable spatial and temporal variability. These include climate, drainage area, relief, geology and availability and calibre of sediment, as well as the vegetational and cultural histories of individual catchments. These considerations preclude simplistic generalisations of anthropogenic-related alluviation, especially the pattern of valley-floor sedimentation induced by deforestation and cultivation (Burrin and Scaife 1988). However, the broad synchrony of Holocene fluvial episodes in catchments with very different land-use histories (particularly in the last 5000 years) suggests that climate may have more directly controlled sedimentation and erosion in British rivers than has hitherto been believed (Macklin and Lewin, in press).

It can therefore be appreciated that integration of Post-glacial valley floor histories and archaeological records requires the identification of anthropogenic-related sedimentation as well as factors that may be specific to particular river catchments or reaches. Geomorphological, sedimentological and mineralogical analyses of valley fill deposits are crucial in this respect in that they can facilitate identification of depositional environments and sediment sources, while pollen analyses and geochronometric dating (^{14}C, TL) allow sedimentary accumulations and associated archaeological material to be placed within chronological frameworks of Holocene climate and vegetational change.

Below we demonstrate this approach through investigations of alluvial valley floors in contrasting upland and lowland settings in north Northumberland and Tyne and Wear. Preliminary results of these studies are presented alongside schematic models of valley floor landform, sediment and archaeological associations evident in north east river valleys that illustrate the environmental contexts and potential sources of 'off-site' archaeological information.

River Valley Archaeology in the Northumberland Uplands

In Northumberland, most upland river valleys (Figure 46) during the Holocene have been characterised by high rates of lateral channel migration and a tendency towards progressive incision of their valley floors. This has usually resulted in the truncation and reworking of Holocene and earlier alluvial fills, associated occupation surfaces and artefactual material. Valley reaches that have experienced particularly high rates of channel reworking may retain little material of archaeological interest on their valley floors. However, earlier Holocene and Late Pleistocene terraces were often utilised for settlement and successive overprinting of cultural landscapes has commonly resulted in a mixed-age assemblage of artefacts, cropmarks and earthworks on their surfaces. For example, in the Milfield Basin, Northumberland, successive phases of Neolithic, Anglo-British and Anglo-Saxon settlement on Late Pleistocene sand and gravel outwash terraces are visible as cropmarks (Hope-Taylor 1977, Miket 1981, Harding 1981, Gates and O'Brien 1988).

Reworking and burial of archaeological features has also occurred on alluvial fans and hillslopes characterised by high rates of Holocene colluviation, while on many valley interfluves archaeological material often lies beneath blanket peat. Thus during the Holocene alluvial, colluvial and paludification processes have resulted in the selective preservation of archaeological sites and landscapes in upland river valleys. Buried soils and organic channel infills, incorporated within alluvial and colluvial sedimentary sequences, do however constititute potentially valuable sources of palaeoenvironmental information.

Small catchments in upland Britain, with comparatively high basin relief and unit stream powers (together with erodable glacially-inherited sediment), frequently respond in a very sensitive manner to rare, high magnitude storm events and localised land-use changes (e.g. Harvey *et al* 1981, Carling 1986, Richards *et al* 1987). The Coe Burn catchment on Callaly Moor, near Rothbury, Northumberland, (Figure 47) has been shown to be typical in this respect (Macklin *et al* in press), where off-site geoarchaeological investigations have been integrated with three seasons of intensive archaeological field survey and excavation prior to afforestation (Nieke 1987a and 1987b, Cowley and Edwards 1988).

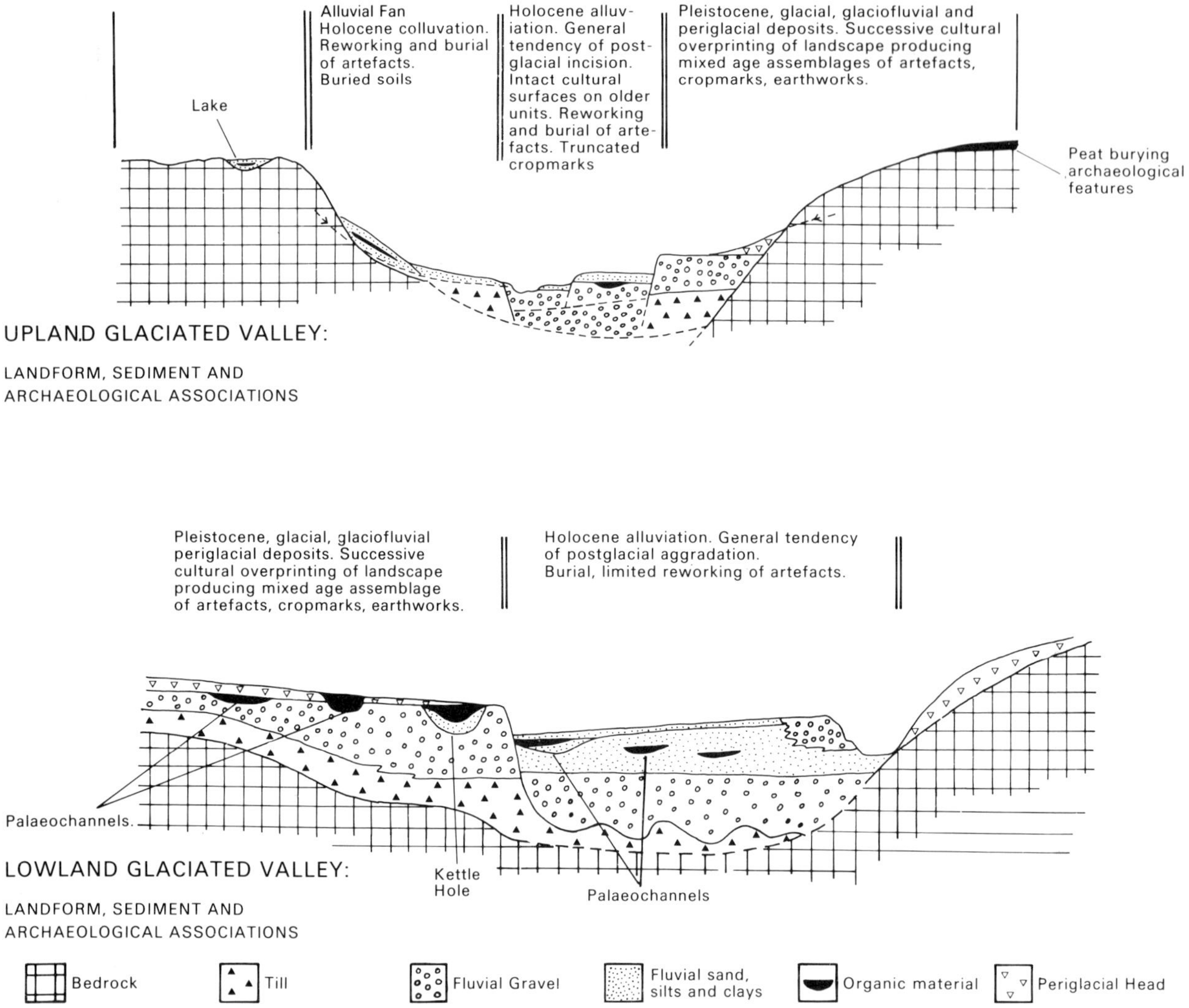

***Figure 46** Schematic cross-sections of upland and lowland glaciated valleys in N E England showing landform, sediment and archaeological associations.*

Callaly Moor lies between 200 - 260m OD and is presently covered with heather and bracken. The moor is underlain by Fell Sandstone with thin coal seams associated with the Scremerstone Coal Group and is dissected by a series of deeply incised gullies that drain into the Coe Burn, a tributary of the River Aln. Field survey has demonstrated evidence of Bronze Age settlement, medieval and post-medieval farmsteads with accompanying field systems, and a large number of bell-pits associated with coal-extraction (Figure 47). Excavation of a stream-bank section in an un-named headwater tributary of the Coe Burn on the western edge of the survey area revealed an anthropogenically-produced deposit of burnt stone, charcoal and ash cut into soliflucted till that underlies the floor of the gully (Figure 48). This feature is ^{14}C dated to the early second millenium bc (3920 ± 70 yr B.P., BETA-28982) and is buried by a 2m sequence of intercalating alluvial and colluvial sediments, separated by peaty soils, overlain by spoil dumped from a bell-pit located on the side of the valley above the section.

Selected physical and chemical properties of these sediments and soil units, together with their probable age and vegetation cover present in the catchment contemporary with their formation, are summarised in Table 1. Two discrete episodes of Holocene alluviation are present, each followed by channel incision and soil development on the valley floor. The oldest alluvial unit is characterised by high values of frequency-dependent (Xq%) magnetic susceptibility which suggests a significant proportion of this sediment was derived from topsoil erosion (Oldfield 1983, Dearing *et al* 1985). Cereals and plant types characteristic of anthropogenic disturbance are also abundant and radiocarbon dating (3920 ± 70 yr B.P., BETA-28982, and 2540 ± 110 yr B.P., BETA-29517) indicates that this phase of alluviation was contemporary with, and probably related to, arable cultivation in the catchment during the Bronze Age. It is interesting to note, however, that although Bronze Age settlement is evident on Callaly Moor, there are no surviving remains in the peripheral tributary catchment where valley fill sediments were investigated.

Pollen analysis of the younger alluvial unit indicates deposition under locally wetter conditions during a period with very little agricultural activity. Lower frequency-dependent magnetic susceptibilities and higher minerogenic concentrations suggests an influx of less-weathered bedrock that is related, at least in part, to mining spoil generated from numerous bell-pits at the head of the valley (Figure 47). Pollen and sediment analyses indicate that catchment destabilisation and associated valley floor sedimentation was induced under a rather different hydrological regime and land-use pattern than in the Bronze Age. This phase of sedimentation appears to have occurred after agricultural contraction on Callaly Moor during the post-medieval period. Since this time, the stream incised through its valley fill in response to the cessation of sediment generated by mining and agricultural activities, and possibly because of an increase in the frequency and magnitude of floods during the Little Ice Age.

Sedimentological and palaeoecological investigations in this small upland catchment, combined with archaeological field survey, has facilitated quite detailed local environmental and land-use reconstructions and clearly demonstrates the value of geoarchaeological methods in elucidating culture-landscape interrelationships.

Archaeology and the River Environment in the Lower Tyne Valley

In contrast to small river basins, short-term but high-intensity climatic and/or anthropogenic-induced erosion/sedimentation events are usually poorly recorded in larger river systems. Holocene alluviation in the lower reaches of large catchments in north east England has in general been controlled principally by longer-term Holocene climate change and also catchment-wide land-use transformations during prehistoric and historic times.

River aggradation near the coast, partly in response to Holocene sea-level rise, has led to the development of laterally extensive and thick Post-glacial alluvial sequences which frequently incorporate organic material. Investigations of these deposits in the lower

Tyne valley (Macklin and Passmore 1988) has been facilitated by the recent construction of the Newcastle Western By-Pass, which traverses one of the largest alluvial basins in the lower reaches of the River Tyne near Blaydon (Figure 49). Sedimentary logs obtained from geotechnical surveys have allowed the pre-Holocene and Post-glacial sedimentary sequences to be defined in detail along a 1700m transect across the valley floor (Figure 50), while two specially-commissioned sediment cores were extracted respectively at Blaydon Haughs and Shibdon Pond (Figures 49 and 50) for detailed palaeoenvironmental analyses.

Holocene alluvial deposits that overlie Late Pleistocene glaciofluvial sandy gravels, can be laterally differentiated into relatively coarse (near-channel) and fine-grained (floodbasin) alluvial sequences (Figure 50). Selected physical and chemical properties of sediments from both lithofacies are presented in Tables 2 and 3, together with associated pollen assemblages and ages of sedimentary units.

Rising alder counts in the lower peaty silty clays (SP 4c) at Shibdon Pond suggest these sediments were deposited under estuarine conditions, probably during the Boreal period and the Boreal-Atlantic transition (c.9500 - 7000 BP) (Moore and Webb 1978). In common with many other coastal sites in England and Wales, this characteristic expansion of alder was most likely related to higher ground-water levels associated with the rapid inundation of lower reaches of the Tyne valley by the rising Post-glacial sea (e.g. Kidson and Heyworth 1982, Smith 1984).

During later Mesolithic and early Neolithic times there was considerable alluviation at Shibdon Pond with deposition of over 4.5m of peaty clays (SP 3) and peaty silty clays (SP 4a - 4b). Pollen assemblages in these units, ^{14}C dated to between 4800 ± 90 yr B.P. (BETA-30346) and 6420 ± 170 yr B.P. (BETA-30348), indicate the local floodplain environment continued to be dominated by alder. A temporary decline in alder counts and rise in Gramineae between 6250 ± 100 yr B.P. (BETA-30347) and 6420 ± 170 yr B.P. suggests an episode of small scale disturbance, possibly associated with localised late Mesolithic woodland clearance. This is consistent with archaeological records of late Mesolithic activity in the surrounding area (Burgess 1984, Miket 1984, Weyman 1984).

There appears to be a major hiatus separating the peaty clay at 3m, dated to ca.4800 BP, and overlying silty clays (SP 2) that were most probably deposited after the Bronze Age. High levels of cereals and other herb species, including Gramineae and *Plantago*, as well as comparatively low arboreal pollen counts are found in the lower part (SP 2b) of this latter unit. Similar pollen assemblages are also present in inorganic silty sands (BH 4) and laminated sands, silts and clays (BH 3) at Blaydon Haughs. Deposition of all three units (SP 2, BH 3 and 4) appears to have coincided with anthropogenic disturbance of forest and soil cover in the Tyne basin that began on a small scale in the Bronze Age, with major deforestation during and after the late Iron Age and Romano-British periods (Davies and Turner 1979, Turner 1983, Higham 1986, Topping 1989). The discovery of

CALLALY MOOR SECTION

Unit	Thickness(m)	Age	Na mg kg^{-1}	K mg kg^{-1}	χq %	Pollen Assemblages	Environmental and Archaeological Linkages
Spoil	0.3 m	Post-medieval	2750	19,500	0.24		Episodic dumping of mining spoil from bell-pit above section, punctuated by stream reworking and soil development
Alluvium	0.05 m	"	2750	19,500	--		
Iron Pan	0.05 m	"	2500	14,500	0.69		
Spoil	0.25 m	"	7500	20,000	0.20		
Buried Soil	0.05 m	"	4000	8,000	4.74		Soil development following cessation of upstream coal mining
Alluvium	0.3 m	"	4000	14,500	2.57	Sphagnum - Calluna - Corylus - Gramineae	Valley floor alluviation related to upstream coal mining in the catchment. Wetter local environment with some limited (pastoral) agricultural activity.
— HIATUS —							
Buried Soil	0.25 m	2540 +/- 110 yr BP (at depth 1.1 m)	4000	6,250	1.57	(upper 10cm) Calluna Gramineae - Corylus Alnus - Cyperaceae (lower 15cm) Gramineae -Calluna - Corylus - Alnus - Cyperaceae	Soil development following decline of cereal cultivation in the catchment. Increasingly wetter local environment towards the top of the unit
Alluvium	0.35 m	Bronze Age	2250	8,000	7.72	Gramineae (including cereals) - Calluna - Corylus - Alnus - Quercus - Betula - Plantago	Valley floor alluviation associated with Bronze Age deforestation and cereal cultivation in the catchment
Burnt Stone, Charcoal and Ash	0.2 m	3920 +/- 70 yr BP					Anthropogenic (Late Neolithic - Early Bronze Age) cut and fill feature, probably a "burnt-mound".
— HIATUS —							
Soliflucted Till	Unexcavated	Late-glacial					Periglacial slope deposits
END OF SECTION							

Table 1 *Summary of the physical and chemical properties, ages, pollen assemblages and environmental/archaeological interpretation of Callaly Moor valley fill sediments.*

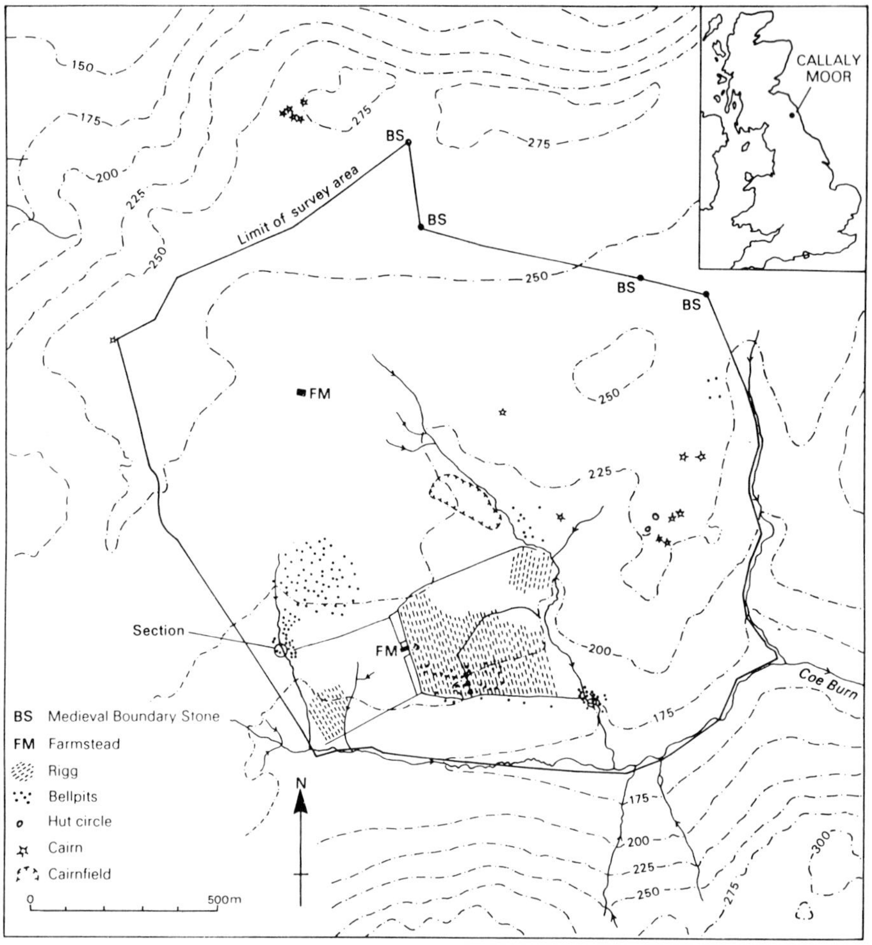

Figure 47 *Map of Callaly Moor, Northumberland, showing field survey area, surviving archaelogy and location of study section.*

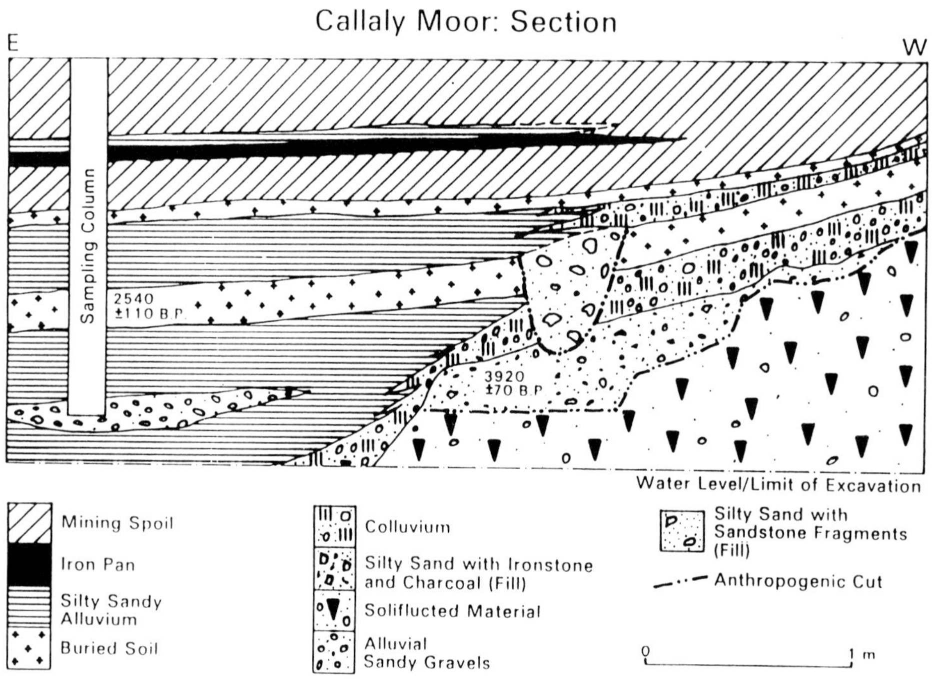

Figure 48 *Callaly Moor valley fill sediments.*

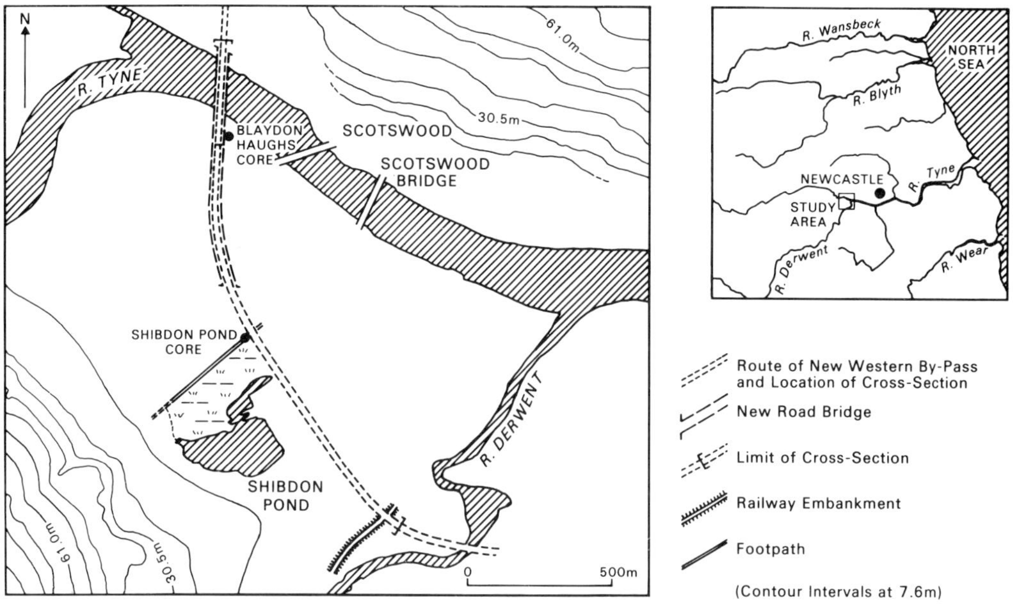

***Figure 49** Map showing valley floor of lower Tyne at Blaydon, route of Western By-Pass and location of Blaydon Haughs and Shibdon Pond boreholes.*

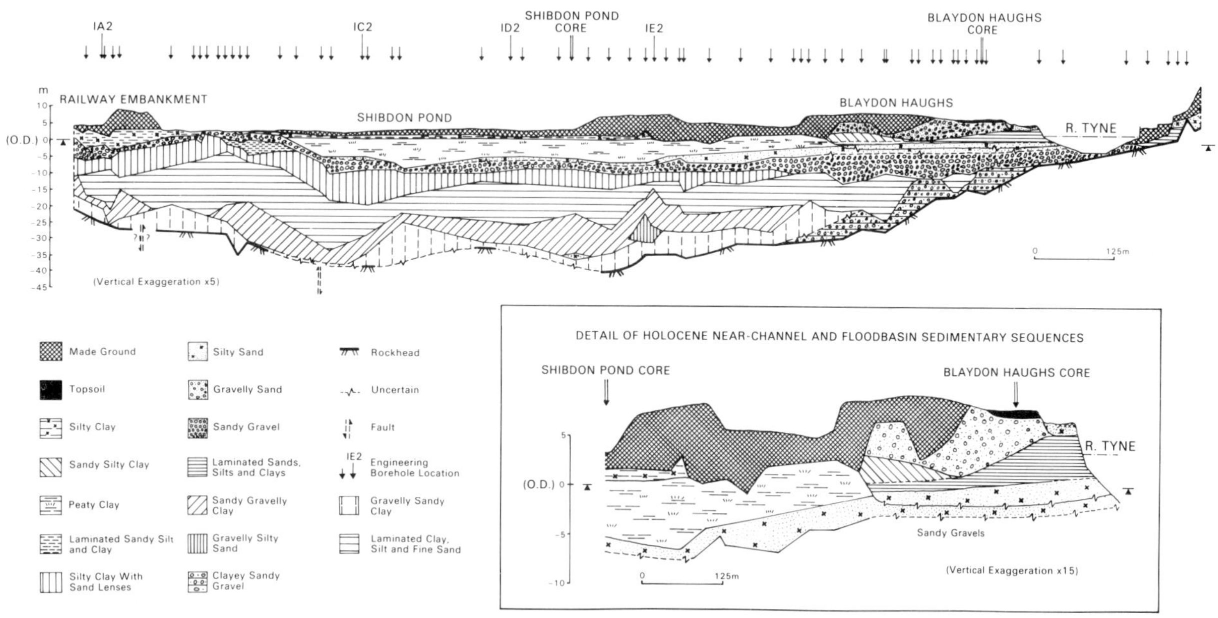

***Figure 50** Cross-section showing Quaternary sedimentary sequence beneath Newcastle Western By-Pass at Blaydon.*

an Iron Age/Romano-British wooden wheel overlying gravels and buried beneath 4.9m of sands and clays on the south bank of the Tyne at Ryton, some 6km upstream of Blaydon Haughs (Piggott 1949), indicates a major phase of fine sediment aggradation began in the lower reaches of the Tyne sometime during and after the Iron Age and Romano-British periods.

Comparatively high concentrations of cadmium and lead are present in gravelly sands that were formerly part of a mid-channel or point bar at Blaydon Haughs (BH 2, Figure 50). Contamination of this unit by heavy metals indicates that these coarser sediments were deposited sometime after large-scale non-ferrous industrial and mining operations were first established in the Tyne catchment during medieval times (Raistrick and Jennings 1965). Relatively low concentrations of zinc in these gravelly sands and the overlying modern floodplain topsoil (BH 1) indicate, however, that alluviation on the valley floor at Blaydon Haughs had ceased by 1880 when the hey-day of zinc mining began in the northern Pennines (Raistrick and Jennings 1965). From the end of the nineteenth century, following dredging and straightening of the channel, the bed of the river has been lowered by between 2-3m, significantly reducing the frequency of flooding in the lower Tyne valley.

Conclusions

Holocene river valleys in north east England offer considerable opportunities for 'off-site' palaeoenvironmental reconstruction that can be accessed through geoarchaeological investigations. Our studies have demonstrated distinct contrasts in the pattern of Holocene alluviation in upland and lowland settings for it is evident that Post-glacial sedimentation in the lower Tyne, in contrast to the upland Coe Burn catchment, has been conditioned by environmental changes of regional significance that reflect long-term climatic shifts as well as catchment-wide deforestation, cultivation and metal-mining activities. These regional patterns of Post-glacial valley floor development have strongly influenced the utilisation of floodplain environments by prehistoric and historic cultures, and have also resulted in the selective preservation of archaeological sites and valley floor landforms. The nature of environmental and archaeological linkages that can be derived from the study of Holocene alluvial deposits is therefore clearly dependent on basin size as well as local rates and styles of alluvial sedimentation.

BLAYDON HAUGHS CORE

Unit (code)	Thickness (m)	Age	Pb	Zn	Pollen Assemblages	Environmental and Archaeological Linkages
			mg kg^{-1} (max concentration)			
Topsoil (BH1)	0.5 m	20th C	200	110		Modern floodplain, waste ground
Gravelly Sands (BH2)	3.5 m	17th-19th C	700	160		Channel bar sediments.
HIATUS						
Laminated Sands, Silts and Clays (BH3)	2.5 m	Late Iron Age-17th C	150	120	Gramineae (including cereals) Corylus - Betula - Alnus - Cyperaceae - Calluna - Sphagnum - Plantago	Overbank alluvial sediments associated with late prehistoric and historic deforestation and cultivation in the Tyne catchment. Laterally equivalent to SP2b
HIATUS?						
Silty Sands (BH4)	2.1 m		100	120		
HIATUS						
Silty Sands (BH5)	1.4 m		50	80		Early Holocene river sediments Laterally equivalent to SP5
END OF CORE						

Table 2 Summary of the physical and chemical properties, ages, pollen assemblages and environmental/archaeological interpretation of Blaydon Haughs core sediment units.

SHIBDON POND CORE

Unit (code)	Thickness (m)	Age	Pb (mg kg^{-1}, max concentration)	Zn (mg kg^{-1}, max concentration)	Pollen Assemblages	Environmental and Archaeological Linkages
Made Ground (SP1)	2.0 m	20th C	800	120		Modern landfill and footpath
Inorganic Silty Clays (SP2a)	0.4 m	Late Iron Age - 17th C	80	110	Gramineae - Corylus - Alnus Quercus - Plantago - Calluna	Overbank alluvial sediments associated with late prehistoric and historic deforestation and cultivation in the Tyne catchment. Laterally equivalent to BH3 and 4
(SP2b)	0.5 m		50	140	Gramineae (including cereals) Corylus - Alnus - Quercus Plantago	
HIATUS						
Very Peaty Clay (SP3)	0.3 m	4800 +/- 90 yr BP (at depth 3.13 m)	75	100	Alnus	Floodplain alder woodland, closed canopy
Peaty Silty Clays (SP4a)	1.75 m		50	130		
(SP4b)	2.5 m	6250 +/- 100 yr BP (at depth 4.9 m) 6420 +/- 170 yr BP (at depth 7.65 m)	75	160	Alnus - Corylus - Quercus Pinus - Ulmus - Gramineae	Floodplain alder woodland, temporary local clearance event between 6 and 6.5 m, possibly associated with Late Mesolithic anthropogenic disturbance on the floodplain.
(SP4c)	1.0 m		50	160	Pinus - Alnus - Quercus Corylus	Establishment of floodplain alder woodland, probably related to rising post-glacial sea levels. Laterally equivalent to BH4
Clayey Silty Sand (SP5)	0.1 m		50	100		Early Holocene river sediments Laterally equivalent to BH5

END OF CORE

Table 3 *Summary of the physical and chemical properties, ages, pollen assemblages and environmental/archaeological interpretation of Shibdon Pond core sediment units.*

References

Bell M., 1981, 'Valley sediments and environmental change', In M. Jones and G. Dimbleby (eds.) *The Environment of Man: The Iron Age to the Anglo-Saxon period*, British Archaeological Reports **87**, 75-90.

Bell M., 1982, 'The effects of land-use and climate on valley sedimentation', In A.F. Harding (ed.) *Climatic Change in Later Prehistory*, Edinburgh University Press, 127-142.

Brown A.G., 1983, 'Floodplain deposits and accelerated sedimentation in the lower Severn basin', In K.J. Gregory (ed.) *Background to Palaeohydrology*, London, Wiley & Sons, 375-397.

Brown A.G. and Barber K.E., 1985, 'Late Holocene paleoecology and sedimentary history of a small lowland catchment in central England', *Quaternary Research* **24**, 87-102.

Burgess C.B., 1984, 'The Prehistoric Settlement of Northumberland: A Speculative Survey', In C.B. Burgess and R. Miket (eds.) *Between and Beyond the Walls: Essays on the Prehistory and History of North Britain in Honour of George Jobey*, Edinburgh, John Donald, 126-175.

Burrin P.J. and Scaife R.G., 1984, 'Aspects of Holocene valley sedimentation and floodplain development in southern England', *Proceedings of the Geological Association* **95** (1), 81-96.

Burrin P.J. and Scaife R.G., 1988, 'Environmental Thresholds, Catastrophe Theory and Landscape Sensitivity: Their Relevance to the Impact of Man on Valley Alluviations', In J.L. Bintliff, D.A. Davidson and E.G. Grant (eds.) *Conceptual Issues in Environmental Archaeology*, Edinburgh Univ. Press, 211-232.

Butzer K.W., 1982, *Archaeology as Human Ecology*, Cambridge Univ. Press.

Carling P.A., 1986, 'The Noon Hill flash floods; July 17th, 1983. Hydrological and geomorphological aspects of a major formative event in an upland landscape', *Trans. Inst. British Geographers* N.S. **11**, 105-118.

Cowley D.C. and Edwards D.N., 1988, *Callaly Moor Field Survey and Excavation, Phase Three, 1988*, Archaeological Unit for North East England.

Davies G. and Turner J., 1979, 'Pollen Diagrams from Northumberland', *New Phytologist* **82**, 783-804.

Dearing J.A., Maher B.A. and Oldfield F., 1985, 'Geomorphological linkages between soils and sediments: the role of magnetic measurements', In Richards K.S., Arnett R.R. and Ellis S. (eds.) *Geomorphology and Soils*, London, George Allen and Unwin, 245-266.

Gates T. and O'Brien C.F., 1988, 'Cropmarks at Milfield and New Bewick and the recognition of Grubenhauser in Northumberland', *Archaeologia Aeliana* **5** 16, 1-9.

Harding A., 1981, 'Excavations in the prehistoric ritual complex near Milfield, Northumberland', *Proceedings of the Prehistoric Society* **46**, 87-135.

Harvey A.M., Oldfield F. and Pearson G.W., 1981, 'Dating of Post-Glacial Landforms in the Central Howgills', *Earth Surface Processes and Landforms* **6**, 401-412.

Harvey A.M. and Renwick W.H., 1987, 'Holocene alluvial fan and terrace formation in the Bowland Fells, Northwest England', *Earth Surface Processes and Landforms* **12**, 249-57.

Higham N., 1986, *The Northern Counties to AD 1000*, London, Longman.

Hope-Taylor B., 1977, *Yeavering: An Anglo-British Centre of Early Northumbria*, HMSO.

Kidson C. and Heyworth A., 1982, 'Sea level changes in south-west England and Wales', *Proceedings of the Geological Association* **93** (1), 91-111.

Lewin J. and Macklin M.G., 1987, 'Metal mining and floodplain sedimentation in Britain', in V. Gardiner (ed.) *International Geomorphology 1986,* part 1, Chichester, Wiley, 1009-1027.

Macklin M.G. and Lewin J., 1986, 'Terraced fills of Pleistocene and Holocene age in the Rheidol Valley, Wales', *Journal of Quaternary Science* **1** (1), 21-34.

Macklin M.G. and Lewin J., in press, 'Holocene river alluviation in Britain', *Proceedings of the Second International Geomorphology Conference, Frankfurt.*

Macklin M.G. and Passmore D.G., 1988, 'Late Quaternary Sedimentation in the Lower Tyne Valley, North East England', *Department of Geography seminar paper no.55, University of Newcastle upon Tyne.*

Macklin M.G., Passmore D.G., Stevenson A.C., Cowley D.C., Edwards D.N. and O'Brien C.F., in press, 'Holocene alluviation and land-use change on Callaly Moor, Northumberland, U.K.', *Journal of Quaternary Science.*

Miket R., 1981, 'Pit alignments in the Milfield Basin and the excavation of Ewart 1, *Proceedings of the Prehistoric Society* **47**, 137-46.

Miket R., 1984, *The Prehistory of Tyne and Wear*, Northumberland Archaeological Group.

Moore P.D. and Webb J.A., 1978, *An Illustrated Guide to Pollen Analysis*, London, Hodder and Stoughton.

Needham S.P., 1985, 'Neolithic and Bronze Age settlement on the buried floodplains of Runnymede', *Oxford Journal of Archaeology* **4**, 125-27.

Nieke M., 1987a, *Callaly Moor Field Survey 1987*, Archaeological Unit for North East England, University of Newcastle upon Tyne.

Nieke M., 1987b, *Callaly Moor Field Survey, Phase Two, Autumn 1987*, Archaeological Unit for North East England, University of Newcastle upon Tyne.

Oldfield F., 1983, 'The role of magnetic studies in palaeohydrology', in K.J. Gregory (ed.) *Background to Palaeohydrology*, London, John Wiley & Sons, 141-165.

Piggott S., 1949, 'A wheel of Iron Age type from Co. Durham', *Proceedings of the Prehistoric Society* **15**, 191.

Raistrick A. and Jennings B., 1965, *A History of Lead Mining in the Pennines*, London, Longmans.

Richards K.S., Peters N.R., Robertson-Rintoul M.S.E. and Switzur V.R, 1987, 'Recent valley sediments in the North York Moors: evidence and interpretation', In V. Gardiner (ed.) *International Geomorphology 1986*, part 1, London, John Wiley and Sons, 869-883.

Robinson M.A. and Lambrick G.H., 1984, 'Holocene alluviation and hydrology in the upper Thames basin', *Nature* **308**, 809-14.

Shotton F.W., 1978, 'Archaeological inferences from the study of alluvium in the lower Severn and Avon valleys', in S. Limbrey and J.G. Evans (eds.) *The Effect of Man on the Landscape: The Lowland Zone*, CBA Research Report 21, 27-32.

Smith A.G., 1984, 'Newferry and the Boreal-Atlantic transition', *New Phytologist* **98**, 35-55.

Thornes J.B., 1987, 'The Paleo-ecology of Erosion', in J.M. Wagstaff (ed.) *Landscape and Culture*, Oxford, Basil Blackwell, 37-55.

Topping P., 1989, 'Early Cultivation in Northumberland and the Borders', *Proceedings of the Prehistoric Society* **55**, 161-179.

Turner J., 1983, 'Some pollen evidence for the environment of North Britain 1000 BC to AD 1000', in J.C. Chapman and H.C. Mytum (eds.) *Settlement in North Britain 1000 BC - AD 1000*, Oxford, British Archaeological Reports **118**, 3-12.

Weyman J., 1984, 'The Mesolithic in North-East England', in C.B. Burgess and R. Miket (eds.) *Between and Beyond the Walls: Essays on the Prehistory and History of North Britain in honour of George Jobey*, Edinburgh, John Donald, 38-51.

Part II

New Techniques and Applications in Geoprospection

Phosphate Survey : Method and Meaning

Rick Walker

Abstract

Phosphate survey is a well known archaeological tool; however, it is little understood in terms of the origins of anomalies, and in interpretation of results. This paper suggests that anomalies develop only in circumstances where organic matter is preserved in ***buried*** *archaeological contexts, and that surface phosphate anomalies reflect underground features containing organic matter to a remarkable degree, in spite of superficial application of phosphate rich materials. It is suggested that the chemistry of buried humic soils should be examined as a next step in establishing the true usefulness of phosphate survey.*

Introduction

Phosphate survey has been employed by archaeologists since the early 20th century, and was first studied in detail by Arrhenius in 1929. As a survey method, it is well known to archaeologists, yet there remain many facets of methodology and interpretation that are obscure. The survey work on which this paper is based has been aimed at the clarification of some of the more obvious aspects that are at present unclear, and that are relevant to the use of phosphate survey as an archaeological tool.

Three areas have been concentrated on; firstly, the extent to which surface phosphate anomalies reflect the presence of underground archaeological features has not been adequately examined. It has been assumed previously that phosphate anomalies on the surface are subject to 'noise' from modern phosphate deposition such as faeces and fertiliser and that therefore only an (unknown) proportion of anomalies found are actually representative of archaeological remains; this paper suggests that this is not a serious problem, and that the assumption involved appears not to be valid.

Secondly, this work has examined the manner in which phosphate anomalies are developed; again, earlier understandings seem to have been seriously flawed, and a number of theoretical models are proposed here for the mechanisms involved. A certain amount of work is planned to test these models, but this area is one more appropriate to soil science and analytical chemistry, and it is unlikely that any firm conclusions can be achieved in the current work.

Thirdly, a considerable amount of effort has been given to the implications of the two areas above for the methodology of phosphate survey and to the applicability of phosphate survey within archaeology.

Brief Literature Survey

Phosphate survey was developed as a large scale survey tool; it has mainly been used to locate and define settlements (Hamond 1983 and others), and has rarely been used as an intra-site survey tool; however, Conway (1983) examined a Romano British hut circle, with interesting results, and much of the work on which this paper is based has been within sites.

In the past numerous chemical methods have been used, including the imprecise 'Eidt' test (see Hamond 1983), and some using rather dangerous chemicals (see Conway 1983); all the work by the writer has employed the method detailed in Craddock (1985). This method is cheap, safe, amenable to field use, and very reliable. The relative merits of different methods will not be discussed here as the addition of further unknown factors by the use of different analytical methods seemed inappropriate in a situation already undesirably obscure. None of the details of methodological procedures designed to improve accuracy of results will be discussed here.

It has been broadly assumed in the past that high levels of phosphate imply that waste (domestic rubbish or animal excrement) was deposited at that point sometime in the past. Hamond (1985) goes to some detail in determining the relative amounts of phosphates deposited by various animals and by various methods. The broad conclusion is that urine is the largest contributor of phosphate by far, but he appears to ignore the fact that faeces and especially bone are subject to rather different post-depositional processes.

Wild (1950) does however discuss the manner in which phosphates are retained in the soil, and it has been shown that phosphate deposited on the surface is rapidly converted to iron and aluminium orthophosphates (except on calcareous soils, where calcium orthophosphate is the main salt produced). These phosphates are highly insoluble in normal soils, but are relatively soluble in acidic solutions, and this fact forms the basis of all the analytical methods in use.

Another important aspect of Wild's 1950 paper was that different soils can absorb different amounts of phosphate; fine humic soils are generally most retentive; gravels and clays are least retentive. This fact is important both in explaining the wide range of 'background' levels that occur on different sites, and in present ideas on the development of phosphate anomalies (see below).

The effects of leaching and of plant action lead to a steep gradient in the levels of phosphate with depth. The maximum levels are found near the surface, where plant litter and growth activity is at a peak, and decays downwards in the area where plant roots are most active in extracting the phosphorus they need from the soil. Below the level of root action phosphate levels continue to decay until the water table is reached because of leaching by ground water.

Having said that the phosphates are highly insoluble, they nevertheless *do* leach from the soil both by water action and by plant activity. If the plants are harvested the phosphate they contain will be lost, but the loss due to water transport continues in areas where harvesting does not occur. The loss by leaching seems to be about 3% per year under normal circumstances (Hamond 1985).

This last factor has not been incorporated adequately into the understanding of phosphate survey; it has been broadly assumed that phosphate anomalies are generated by features containing high levels of phosphate when buried; while this may have been so for certain features, it is perfectly clear that any soluble phosphate in the deposit will decay away completely in just a few decades, let alone over millenia. The fact that phosphate anomalies *do* correlate with archaeological features indicates a major gap in our understanding of the origin of the anomalies.

Sites Examined

The sites from which samples have been analysed by the writer in the course of present work are listed below, with a summary of the results and dot-density diagrams where relevant.

1. Trethallen Farm, Newquay, Cornwall. Bronze age settlement and iron age cemetery.

This survey comprised two parts.

Part 1 was an examination of about 200 context samples, and was intended partly to gain familiarity with the methodology, and partly to confirm the generally held view that features containing large quantities of organic waste are likely to display high phosphate values. This was broadly confirmed, but it was clear that each category of feature (such as post-hole, pit, hearth layer) had a considerable spread of phosphate levels. It was also found that some dark humic samples had rather low phosphate content and that some pale samples had high phosphate content. In conclusion it was felt that there is no way that phosphate content can be estimated from the type of context, the colour of the fill, or the amount of charcoal present.

Part 2 examined a number of the iron age graves, some with and some without skeletal remains, and a stone built structure probably of bronze age date. In spite of rather crude sampling intervals, the results all confirmed the findings from graves with skeletal remains, that the bodies were located at the north end of the grave. The results from the hut were less easy to interpret; one third of the floor had already been cleared before the sampling was undertaken, and the date was very 'noisy'. However, the smoothed results (Figure 51) do clearly show that the edges of the hut were low in phosphate (contrary to Conway 1983 from his RB hut), and that the highest phosphate levels were found in the area behind the central post relative to the entrance. Provisionally, the results were interpreted as indicating deposition of phosphate rich organic matter in the middle of the hut, around the post. It is assumed that the high level behind the post is due to the relative difficulty of removing rubbish from this area, whereas the edge and the area near the entrance would be relatively easy to clear.

The reasons for the difference in phosphate distribution from Conway's Romano-British hut is unclear; possibly subsequent use by animals occurred in Conway's hut; the hut at Trethallen seems to have been filled with rubble and hillwash after it went out of use.

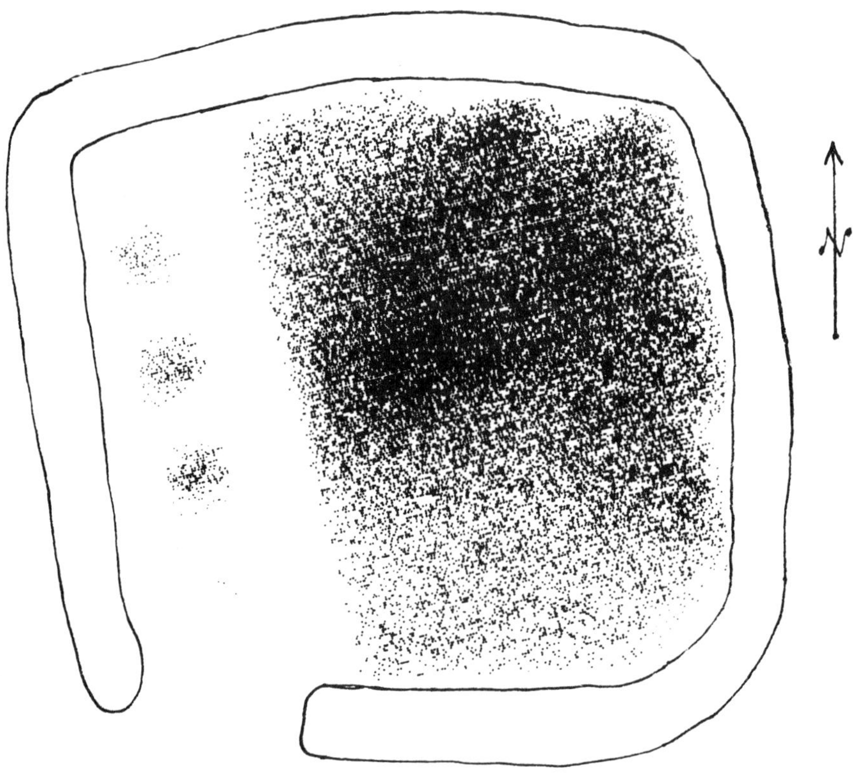

***Figure 51:** Square Bronze Age hut at Trethallen*

2) Reawla Round, Cambourne. Iron age bivallate defended enclosure.

Work on this site involved taking about 1900 samples at 2m intervals over the entire site, including an area that had already been excavated. The area that had been excavated showed some correlation of excavated features with phosphate anomalies, and some anomalies were found that did not correlate with excavated features (see Figures 52 and

53). It is important to note here that the phosphate samples were taken at a level *below* the excavated levels, implying that phosphate anomalies can be detected at any level in the soil; this result confirms Hamond's 1983 suggestion (see below for further discussion).

The rest of the site contained a considerable number of anomalies, but it proved extremely difficult to offer any interpretation for these features as there were no parallels available elsewhere, and because the correlation within the excavated area was not good enough to allow extrapolation to other areas. Nevertheless, the results were very interesting. Circumstantial evidence suggested that most of the anomalies observed were aligned with known earthworks or features noted by magnetometer survey or by the subsequent watching brief. The most difficult anomalies to resolve were also the most obvious, and were inexplicable until a chance comment about the possibility of a third concentric rampart. This possibility could adequately explain the phosphate results, and there were aspects of an earlier magnetometer survey by the AM Lab that supported this hypothesis. In addition, the anomalies found in the excavated area that did not correlate with excavated features could also be explained by a third rampart. Obviously the suggestion that the excavators may have failed to find a major feature on the site cause some acrimony, but there remains the possibility either that it was very deeply buried, or that it was very superficial and had been removed during clearance of topsoil by JCB.

3) Lismore Fields, Buxton, Derbyshire. Neolithic settlement.

This site was on an extremely wet low plateau. The survey involved two phases, firstly an examination of context samples from the post-holes of the structures found. It was clear that certain of the structures had been much more heavily used than others, though it remains unclear whether this was by human or animal activity. Indeed, the results from this site led to some discussion as to whether animal deposits actually appear at all in phosphate surveys; all animal deposits are superficial, and unless their faeces are buried, it seems likely that their presence will not lead to a phosphate anomaly. It is hoped that magnetic susceptibility data will shortly assist the interpretation of the site.

The second phase of the survey was an examination of part of the excavated area at 1 metre intervals. A number of anomalies were found that clearly correlated with excavated Neolithic features, and a number of other anomalies were found that correlated with known modern drainage trenches. The results from this site were very interesting because, firstly, they confirmed that respectable phosphate anomalies can be associated with remains as old as the Neolithic, implying that whatever mechanism causes the anomalies is not necessarily degraded by time. Secondly, a popular path used by dog owners, with consequent deposition of large quantities of dung and urine, did not appear at all in the results. This supports the suggestion from Reawla and other sites (see Potlock below) that the superficial application of large amounts of phosphate, in the form of dung or fertilisers, is not necessarily reflected in the distribution of phosphate in the soil (see paragraph above and below for further discussion).

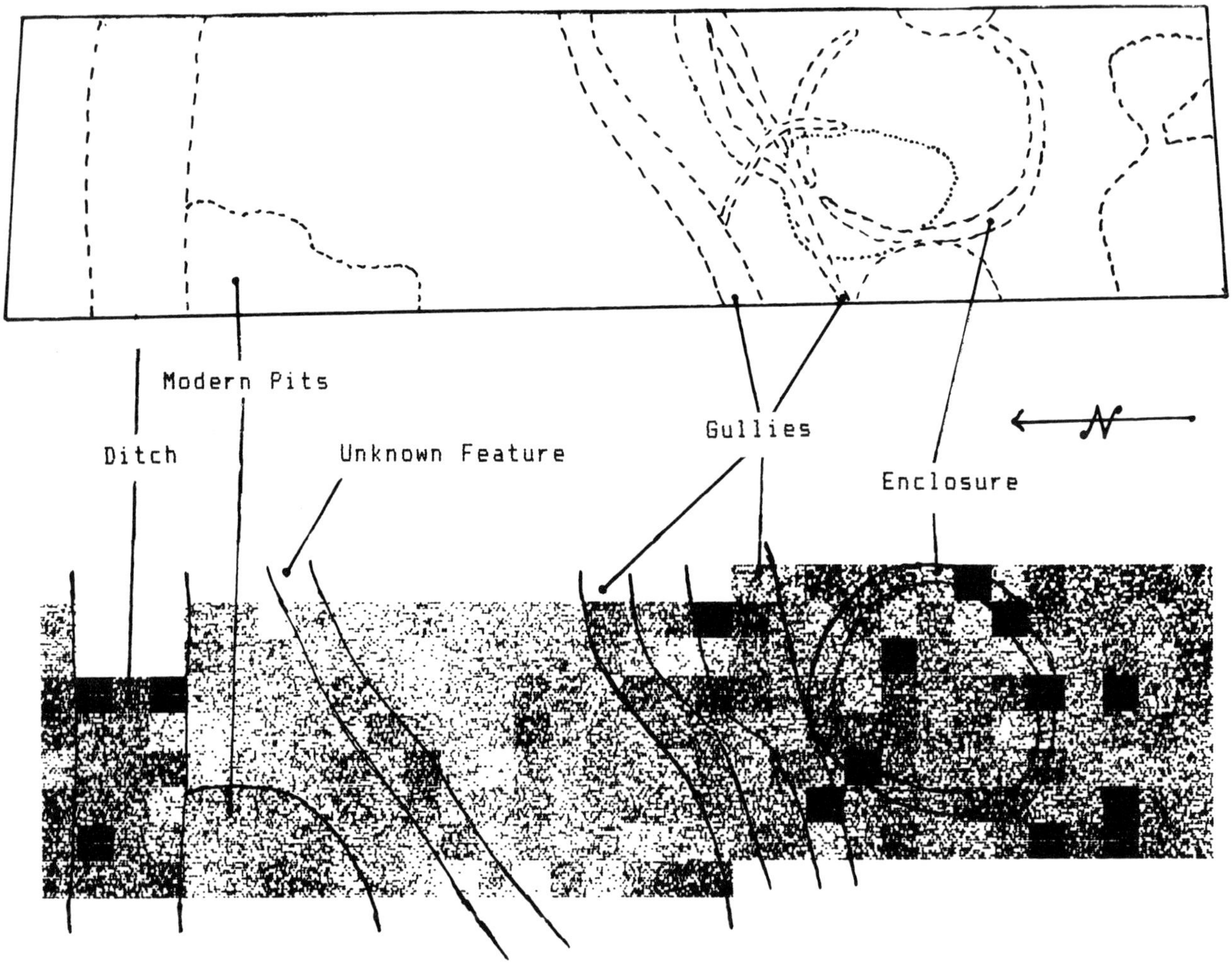

Figures 52 and 53: *Reawla Round; Excavated and Phosphate Features. (Excavated features simplified from site drawing). Scale 1:400*

4) Ribchester Roman Settlement

This site was examined in a small test survey because it allowed two possibilities to be tested. It had been suspected that the good results at Lismore Fields had been partly due to the wetness of the site, allowing good preservation of organic remains, and therefore developing good phosphate anomalies. The water table at Ribchester was very high indeed, only 30 - 50 cm below grass, and therefore a similar situation should occur as at Lismore Fields. The other possibility, suggested as a result of data collected in Greece (C. Buck pers. comm.), which suggested that waterlogging prevents the development of phosphate anomalies, was also testable on this site, as it was frequently heavily waterlogged.

In fact the soil at Ribchester was very high in phosphate, implying firstly that waterlogging in no way prevents the development of phosphate anomalies, and supporting the hypothesis that phosphate anomalies are dependent on the state of preservation of buried organic matter.

The anomalies found at Ribchester were hard to interpret as the areas examined were too small, but there were clear striations across part of the site in a different direction to any excavated remains (possibly due to geological effects or river action), and in a small area examined at 20cm intervals, a possible grid-based series of high phosphate anomalies were observed. The cause of this effect is unclear, but may be involved with a Roman casket cemetery, or (more probably), may be some artefact of ploughing methods (A. Olivier pers. comm.).

5) Potlock, Derbyshire.

This is a multi-phase site identified from aerial photographs. A phosphate survey on a 5m grid was undertaken as a prospection tool in the first instance, to attempt to locate areas warranting detailed examination by other means. The results were extremely noisy, as might be expected from a complex site, but it was possible to identify areas where phosphate levels were relatively high, and to suggest that they may be foci of settlements or other human/animal activity. Little more can be said at present as this work is ongoing, and correlation with the results of resistivity, magnetic susceptibility and magnetometer survey, will be attempted in the near future. Nevertheless, the site is open agricultural land and is no doubt regularly fertilised, confirming suggestions made above that superficial application of phosphate-rich materials do not significantly influence anomaly formation.

Theoretical Discussion

The above results offer a measure of clarification as to what is involved in the development of phosphate anomalies in the soil.

Firstly, the fact that anomalies can be found in contexts as old as the Neolithic confirms very clearly that the mechanism of phosphate retention in the soil has nothing to do with the amount present when the deposits were buried. As mentioned above, leaching and plant action will remove all the phosphate present very quickly (in archaeological terms). Even if Wild (1950) was wildly in error by a suggested 3% per year leaching rate, at an improbable 0.1% per year the phosphate present in a neolithic rubbish heap would have decayed to less than 2% of its original value, quite undetectable against background noise.

Secondly, it is clear that superficial deposition of modern phosphate does not significantly influence the detection of phosphate anomalies. This is rather surprising, as one would expect that excrement or rubbish dumped on the surface would be rapidly absorbed into the soil, leading to a high phosphate area or an apparent anomaly. The fact that the dog watering path at Lismore field was not detected, and the fact that most of anomalies at Reawla were probably archaeological in origin rather than due to modern rubbish and dog excrement leads to the strong suspicion, supported by work on other phosphate surveys, that whatever mechanism is responsible for phosphate anomalies, it is not affected by surface deposition to a significant extent.

Thirdly, the possibility that the results at Lismore Fields were so good because of the wetness of the site, supported by the results at Ribchester, indicated the strong possibility that phosphate anomalies are dependent for their development on the preservation of organic remains. The relatively low levels of phosphate found at Potlock (in spite of regular use of fertilisers) may thus be due to poor preservation of organic remains in the soil. This is likely as the site is on sand and gravel and, although the water table is probably not deep, the soil is loose and open and thus sufficiently well-drained to prevent preservation of organic materials. The same applied at Trethallen, which was on well drained sandy soil over slate. The situation at Reawla was slightly different in that the site was on clay, quite wet in winter but very dry in summer, and it is known that clay is generally less retentive of phosphates than loose soils. The low levels found on that site are therefore also in keeping with the poor level of organic preservation in buried but not waterlogged contexts.

To summarise the present theoretical view of phosphate anomalies, it seems probable that the development of anomalies is dependent to a very large extent on preserved organic remains in buried archaeological features, and that phosphate levels will be enhanced in circumstances where preservation is good, such as in waterlogged sites.

If this theoretical possibility is confirmed by future experiments, it has important implications for interpretation of results, suggesting that processes that do not lead to burial of organic waste will not appear in phosphate anomalies. It is therefore far more probable that human activity alone will appear in the results of phosphate surveys, and it is likely that animal activity will not be readily detectable.

It would appear that the mechanism of phosphate anomaly development is that archaeological deposits preferentially retain phosphates passing through the soil, where they are then made available to plants via root action. It is assumed that this retention by archaeological deposits is due to the presence of organic matter, but this need not necesarily be the case. For example, some of the data from Trethallen, with high phosphate levels but apparently low organic content would appear to have some other component which retains phosphates.

This would imply that future investigation of the mechanism of phosphate retention should concentrate on establishing what component of archaeological deposits is responsible for the retention of phosphates. A number of mechanisms are possible, including:-

A) The retention of phosphates is dependent on pH (Wild 1950), and it may be that some archaeological deposits are better at retaining phosphates because the pH is more suitable.

B) Biological activity may be responsible for retention of phosphates. It seems quite probable that the micro-organisms that occupy archaeological deposits would be able to make good use of phosphates, and a mechanism for retention of phosphate would be advantageous, especially in view of the competition with plant roots for minerals (see Wild 1950).

C) It is possible that the presence of modest-to-large quantities of charcoal, which is good adsorbing chemicals on to its surface, may be responsible for retention of phosphates (see Wild 1950).

D) The presence of degraded bone and other materials (some already rich in phosphates), may act as complex structures in which biological and chemical processes may occur, leading to phosphate retention.

E) Other more complex possibilities exist; it may be that the archaeological layers contain trace elements that are very valuable to plants, and that it is in fact plant roots that retain the phosphate, and these resultant anomalies may be linked with features by very tenuous circumstances indeed.

F) It remains possible that some phosphate anomalies are due to the presence of large quantities of bone in the buried features. It is very unlikely that all phosphate anomalies are created in this way.

Practical Considerations

The surveys that form the background to this paper have led to some development of practical methodologies in the field. These points are included partly because they may be useful to future workers in this field.

1. It is crucially important to take samples at a constant depth; the gradient of phosphate level through the soil is such that considerable errors occur if samples are taken at variable depth. The writer now uses a simple auger with a depth stop, to sample about 5 grams at a depth of 13-15cm (below the zone of greatest root action, but still above most of the archaeology). Up to 500 samples can be taken in a day with this method, and site disturbance is minimal.

2. Phosphate anomalies appear to be present vertically throughout the soil section, and although they are smaller in magnitude at greater depth, they nevertheless appear to

retain their horizontal plan. Therefore, it is likely that a sampling program carried out post-excavation will be lower on phosphate level but will display the same anomalies as a surface sample. This may be very advantageous where site interpretation is unclear, and other sources of data are needed.

3. It is commonly assumed that subsoil and topsoil samples are needed as 'controls'. This is quite unnecessary, as high anomalies are visible relative to low phosphate areas of the survey, and due to the gradient through the soil, samples from different levels in the soil are of little use.

4. Sampling should be arranged at an interval capable of finding what is being searched for. This applies to wide-spacings, but is also equally applicable at the other end of the spectrum, as for example at Trethallen, where it was probably a waste of expensive analytical work to sample at 10cm intervals in a hut. The results, not surprisingly, had more relevance to the topography of the floor of the hut than to the overall phosphate variability. At the large end of the survey spectrum, the limit of acceptability is probably 10m intervals, because above that it is impossible to differentiate local 'noise' (always a serious problem in phosphate survey) from archaeological anomalies. It is probable that surveys should not be conducted at much more than 5m intervals (and even this was really too great at Potlock; the data is too coarse to differentiate features on such a complex site).

The writer feels that a 1m interval is probably ideal for large scale intra-site work, with 20cm intervals for very complex small areas. A 2m interval (as used at Reawla) is probably ideal for prospection within fields, and 5m intervals are suitable for landscape survey. Obviously, the final decision as to what survey interval to use will depend as much on finances as on ideals, but there is no doubt that the improvement in results from a smaller interval survey is enormous.

5. As a final point in this section, it should be made absolutely clear that there is no substitute for the most rigorous analysis possible. As Gurney (1985) points out, the method detailed in Craddock (1985) is only semi-quantitative in its aim, but the reduction of experimental errors results, in effect, in the removal of a large component of the 'noise' associated with phosphate survey, and a clearer set of results will be possible with rigorous methods. In the writer's experience, a 'noise' component equal to about 15% of the mean phosphate level will be found on all sites while experimental errors can be reduced to 3% if great care is taken and systematic assessment of the accuracy of the results is undertaken. At present the exact cause of the noise in phosphate data is unclear. Every attempt is made to analyse a representative part of each soil sample, but there seems to be no way to get absolutely accurate results from a single chemical assay. Multiple analyses on single samples have a normal distribution from which the mean can be accurately estimated, but this is far too expensive for normal survey. Phosphate survey is not a method suited to the fainthearted or technically careless.

Future Work

The main thrust of the work so far has been to establish that it is *archaeological* deposits (ie buried matter) that generate phosphate anomalies; this means, as David Jordan (pers. comm.) has vigorously pointed out, that archaeologists must turn to the study of the soil science of *buried* layers if understanding of phosphate survey and other post-depositional processes are to improve. For some reason soil scientists seem to have ignored burial features, and archaeologists have not paid much attention to such matters, apart from those working on the preservation of artefacts and the decay of bodies.

It is felt that the present work has established adequately that it is archaeological deposits alone that cause phosphate anomalies. Further survey work without studies on the mechanisms in action will probably not help much further in improving interpretation.

The hypothetical mechanisms mentioned above are all testable, and no doubt other tests could be carried out, but much of this work is outside the present areas of expertise of the writer (largely though lack of research facilities). Much of this work will have to be left to others, and it is likely that a long term project will be needed that can allow time for plants to grow into, and to modify, artificial archaeological environments.

Applications of Phosphate Surveys

The present work has suggested that phosphate surveys are much more reliable than previously thought in terms of indicating areas of archaeological activity, in particular areas where organic remains have been buried. This satisfactory result is counterbalanced by the clear fact that it is still uncertain how anomalies can realistically be interpreted, and it seems to the writer that phosphate survey will for the time being be limited to two broad types of problems.

Phosphate survey can have very valuable interpretative implications for excavated sites where interpretation of features is unclear. This kind of survey is very valuable as regards the gain in experience of correlation of known features with phosphate anomalies.

The use of phosphate survey in landscape-scale prospection remains entirely valid and well proven. In these circumstances it often does not matter greatly exactly what types of archaeological features are being observed, and interpretation is thus relatively simple.

The only proviso in both these areas of interest is the suggestion that animal deposits may often be hard to detect, and that therefore interpretation of certain sites should take this possibility into account.

It is to be hoped that the considerable potential of phosphate survey as a prospection and interpretative tool will continue to develop; it remains the only practical and cheap large-scale chemical survey method and as such can detect archaeological activity where no other method can. Considerable work will be needed to make interpretation of the results easier, and some faster (and cheaper) analytical system is needed to make phosphate survey economically competitive with geophysical techniques.

Acknowledgements

None of the work on which this paper is based would have been possible without the generous assistance of the directors of all the sites involved. The following people have given permission for details of the surveys to be used here:

Trethallen:	J Nowakowska, Cornwall Archaeological Unit
Reawla:	N Appleton, Cornwall Archaeological Unit
Lismore Fields:	D Garton, Trent & Peak Archaeological Trust
Ribchester:	Adrian Olivier, Lancashire Archaeology Unit
Potlock:	Graeme Guilbert, Trent & Peak Archaeological Trust

References

Arrhenius O., 1929, 'Die Phospatfrage', *Zeitschrift fur Pflanzenernahrung, Dungung und Bodenkunde* **14**, 185-94.

Appleton N., forthcoming, 'Reawla Round Excavation', *Cornish Archaeology.*

Conway J.S., 1983 'An Investigation of Soil Phosphorus Distribution within Occupation Deposits from a Romano British Hut Group', *Journal of Archaeological Science* **10**, 1983, 117-128.

Cook S.F. & Heizer R.F., 1965, *Studies on Chemical Analysis of Archaeological Sites*, University of California, University Press Publications in Anthropology **2**, 1-9.

Craddock P.T., Gurney D., Pryor F. & Hughes M.J., 1985, 'The Application of Phosphate Analysis to the Location and Interpretation of Archaeological Sites', *Archaeological Journal* **142**, 361-376.

Garton D., 1987, 'Buxton, Lismore Fields', *Current Archaeology* **103**, 250-253.

Gurney D., 1985, *Phosphate Analysis of Soils, a Guide for the Field Archaeologist*, Institute of Field Archaeologists Technical Paper 2.

Hamond F.W., 1983, 'Phosphate analysis of Archaeological Sediments', in *Landscape Archaeology in Ireland*, Eds Reeves-Smythe T. & Hamond F.W., BAR British Series **116**, 47-80.

Nowakowska J., forthcoming, 'Trethallen Farm Bronze Age Excavation', *Cornish Archaeology*.

Wild A., 1950 'The Retention of Phosphate by Soil', *Journal of Soil Science* **1**,ii, 221-238.

Tools for the interpretation of soil phosphate data from archaeological surveys

C.E. Buck, W.G.Cavanagh and C.D. Litton

Departments of Archaeology and Mathematics, University of Nottingham, UK.

Abstract

Despite the fact that soil phosphate analysis is one of the better known techniques for the surface surveying of archaeological sites it is not in widespread use. A survey of soil phosphate concentrations can provide information about aspects of the archaeological activity on a site not available to us via any other method. The reasons behind the slow adoption of soil phosphate analysis as a routine method in archaeological field survey are discussed, and improved methods of data analysis, that might make it more viable, are suggested. Using case studies from real archaeological surveys, undertaken with very different aims in mind, we illustrate two Bayesian statistical techniques appropriate to the analysis of soil phosphate data. An advantage of the Bayesian approach is that it enables the researchers to incorporate their expertise and understanding into the analysis. As a result of the statistical analyses enhanced images of the data are produced. The archaeological interpretation of these images is discussed in the light of other archaeological information.

Introduction

By means of techniques such as resistivity survey, magnetic susceptibility, trace element analysis, intensive surface survey, soil phosphate analysis and others, an archaeological landscape which hitherto was invisible is now being revealed. It is no exaggeration to compare the impact of these techniques in the past 20 years with the state of aerial photography in the 1920's and 1930's. The interpretation of this hidden landscape will colour our understanding of the past, will influence excavation strategy and perhaps help with planning and conservation. We note that it is not uncommon for field archaeologists to complain how much easier it is to press for action when a site is shown up by visible markings on an aerial photograph. There is a challenge to those working on the 'hidden landscape' to make it ever clearer and more understandable by the non-expert, by improved methods of display and graphical representation. The parallel with aerial photography, moreover, can be pushed even further. The manuals on aerial photography are full of warnings to the unwary about pitfalls in the interpretation of photographs by the incautious and the uninitiated. Traps have been laid by effects as various as pleistocene ice-wedges and the ring of the local gymkhana. In the study of hidden landscape using geoprospection techniques, we are faced with similar problems of noise and of interpretation which only the knowledge of the expert can resolve.

Of the wide range of geoprospection techniques available to archaeologists soil phosphate analysis is one of the better known, but less commonly used. There is a wealth of direct evidence to show that enhanced levels of soil phosphate commonly (but not always) occur on archaeological sites, and a rich literature attests this (for example Arrhenius 1938, Craddock *et al.* 1985, Dietz 1957, Hamond 1983, Keeley 1981 and Widgren 1983). We have, then, every reason to accept that soil phosphate survey can help to indicate areas of activity by ancient man. The relatively high concentration of phosphate is, of course, a reflection of the former rich organic component in archaeological soils and it has been found that modern activity does not seriously mask the archaeological signal. Chemical analysis of soil samples for phosphate content can be undertaken fairly rapidly with a simple, quantitative method being available for use in the field (Craddock *et al.* 1985). However, difficulties appear to arise in the selection of suitable statistical methods for treatment of data and in the subsequent interpretation. This is probably due largely to the particularly noisy nature of the data. Such problems have led us to investigate possible mathematical and statistical methods which might aid archaeologists in analysing the data, both in the field and back in the laboratory.

There are many ways of analysing and displaying spatial data, ranging from relatively simple methods such as dot density plots, greyscales and colourscales to the mathematically more sophisticated techniques of contouring and image processing. In any situation, the choice of technique must depend upon the nature of the data and the problem posed by the archaeologist. For example, we know that there is considerable variation in the background phosphate level (see Cavanagh *et al.* 1988): techniques such as dot density may depict this variation (which is of little archaeological importance) as well as describing the on-site variation. (Of course background values may be suppressed in the graphical output, but there still remains the problem of the selection of a suitable threshold level). Moreover, the accuracy of the chemical analysis is important. If, as in the two cases described later, the chemical analysis is of the quick, cheap but crude type, then this should be borne in mind at the data analysis stage. Using a sophisticated technique such as contouring on data that are subject to relatively large measurement errors is not at all sensible and any results produced may be misleading or unreliable.

Another crucial point to be considered is that the data analysis can proceed by either of two routes. The first attempts to produce a description or visual image of the observed data. The second postulates a statistical model of the spatial distribution of the phosphate levels and then proceeds, by using statistical techniques, to estimate the parameters of the model and so produce an image. The former approach only describes the data obtained whereas the latter provides a means of investigating the underlying process generating the data. For a discussion of a similar point in another context see Fieller and Flenley (1988).

In simple situations where the noise level is low or, equivalently, where there is little variation in the background level, dot density, contouring and greyscales can be very effective in aiding the interpretation. However the choice of how many levels to display and what these levels should be remains to be made. Usually simple sample statistics, such as mean, standard deviation, quantiles etc of the raw data (perhaps after some simple transformation and/or filtering) are used. However these techniques do suffer from serious deficiencies which severely limit their use in more noisy and complex situations. For example, contouring is susceptible to noisy or anomolous data points and is not very suitable if fairly abrupt changes in level are to be expected; dot density and greyscales assume that any noise in the data is relatively small and so that an observation is an accurate representation of (in our case) the true phosphate level.

For these and other reasons we belive that dot density, greyscales, contouring and similar techniques are inappropriate to the spatial analysis of soil phosphate data arising from field surveys. Moreover, at this stage in the development of phosphate analysis for the investigation of archaeology, since the relationship between the archaeology and chemistry is little understood, it seems inappropriate to try to determine from the data anything more than areas or zones that have roughly the same phosphate level. If we accept this, then the problem becomes one of image segmentation (which should not be confused with image enhancement used in aerial photography) and we have found two methods to be useful, namely (a) the change-point methodology described in Buck *et al.* (1988), (b) the image segmentation algorithm of Besag (1986).

We shall demonstrate these techniques using two particular soil phosphate data sets as case studies. The first of these derives from intensive field survey in Laconia, Greece and is used to investigate the extent of a site previously recognised by surface pottery and tile scatters. The second comes from an excavated horizon within a Bronze Age cairn at Manor Farm, North Lancashire (see Oliver, 1987, for a full description of this site). Both are real sites and have the complexity and difficulty of interpretation which often confront the field archaeologist.

Although similar in nature, with both data sets being collected over a regular grid, the aims of these surveys were quite different and the archaeological questions posed and statistical techniques adopted will reflect this. Our intention in this context is to stress the importance of two fundamental points, 1) the choice of an appropriate sampling scheme and sampling interval and 2) the selection of a statistical method appropriate to the specific aims and strategy of the investigation.

The archaeological context and methodology

In order to demonstrate the statistical methods we have selected data from two contrasting types of archaeological soil phosphate survey.

(a) **Site SP170, Laconia, Greece**

The site with current code name SP170 was located by the Laconia survey (joint directors W.G. Cavanagh and J.H. Crouwel) during intensive field survey in Laconia, Greece. It was originally discovered during fieldwalking and appeared on the surface as a scatter of tile and sherds. The information obtained from the sherds indicates that the site is of the Ottoman period and provides a guide to the location of the original occupation area. Local informants identify this location with Morou a hamlet mentioned in historical sources. Commonly in intensive survey artefacts are used to define the extent of a site, and it was considered useful to compare this with the results from other methods, in particular geophysical prospection and soil phosphate survey. The results here represent one of ten sites thus surveyed.

The aim of the phosphate survey then was simply to establish the extent of the occupation area as defined by enhanced soil phosphate concentration. With this in mind a survey interval of 10m was selected. The area surveyed was defined by placing the highest intensity of sherds at the centre of a 16 by 16 grid; with soil samples being taken at a depth to about 5cm from the surface.

(b) **Manor Farm, Borwick, North Lancashire**

In 1982 Adrian Oliver, under the auspices of the Cumbria and Lancashire Archaeological Unit, directed the excavation of the Bronze Age cairn at Manor Farm in advance of gravel quarrying. Following completion of the excavation, when inhumation and cremation graves had been revealed, the excavation team collected soil samples from the excavated surface. It was hoped that by phosphate analysis of these samples it would be possible to reveal the location of any graves not recoverable through excavation, but still remaining as enhanced phosphate in the soil. With this aim in mind a survey interval of 0.5m was selected. Given the size of the recorded scatterings of bone this seemed a reasonable choice; two or more readings might be expected to fall within any area of deposition. The whole of the area of the cairn was surveyed, with samples being taken at or near the excavated surface.

Chemical Methodology

The soil samples from both surveys were analysed using the same chemical method, that of Craddock *et al* (1985). This method has proved suitable for use in the field because it is relatively quick, uses simple equipment and can, if necessary, be carried out by volunteers: on the Laconia Survey second year chemistry undergraduates. Of course there is a price to pay for simplicity, the results, although quantitative, are imprecise and, in absolute terms, inaccurate. The reproducibility of the method has been found to be 3% (Conway 1983 and Cavanagh *et al.* 1988). In the present uncertain state of knowledge about the origins and movements of phosphate compounds in soil, archaeologists have been prepared to tolerate such imprecisions. The truth of the matter is that the processes leading to the phosphorus compounds that we measure in the soil are poorly understood and post-depositional changes have not been well researched. Summing up the situation in 1976 Proudfoot wrote "the composition of phosphorus compounds, and their mode and rate of formation in soils, are still uncertain..." but "These difficulties need not deter continued use of phosphate analysis in helping to solve archaeological problems. The necessity is for an understanding of the complex nature of the problems involved." Research needs to be conducted here, yet in thirteen years since 1976 this challenge, it appears, has not been taken up. In the absence of such fundamental research archaeologists have continued to use phosphate analysis, but on a broad scale, in the belief that crude but generally valid results will emerge, provided large and carefully managed samples are taken. The alternative is to ignore a source of evidence whose validity is firmly, if empirically, established.

As has been pointed out earlier, the variation within the background levels of soil phosphate is high (Cavanagh *et al.* 1988). Given the relative inaccuracy and imprecision of the chemical analyses, the uncertainty as to what, exactly, is being measured and the variation in the non-archaeological background, it is clear that at present noisy data are unavoidable. It has become obvious to us that understanding and interpretation of the data might be improved through statistical treatment. Whilst some geophysical data might best be displayed through a continuous grey-scale, indicating subtle changes of level, this does not seem appropriate for phosphate readings. The inaccuracies and uncertainties explained above led us to techniques of image segmentation, whereby zones, in this case, of high readings are demarcated within a noisy background, taking care that too much significance is not placed on isolated individual observations. It might be noted here that similar problems of noise also appear to be found in other geochemical data, such as trace metal concentrations investigated as part of the Cambridge and Bradford Boeotian Archaeological and Geographical Expedition (Bintliff *et al.*, forthcoming and this volume).

Statistical methodology

Before carrying out any statistical analysis of the data, we must be clear about the archaeological questions being posed and how these may be translated into a statistical model. Broadly speaking, the archaeologist wishes to divide the survey area into zones of different soil phosphate concentrations. In statistical terms, this is interpreted as dividing the survey area into different zones where, within a zone, the phosphate level is constant apart for some noise. That is in zone i, the mean level is supposed to be μ_i and the variability due to the noise is represented by the standard deviation, s_i, of the phosphate reading above this level. Furthermore, it is supposed that the phosphate readings (perhaps after some suitable transformation) within zone i have a normal distribution, with mean μ_i and standard deviation s_i.

Having defined our statistical model, albeit a simple one, we now describe two methods of data analysis that seem appropriate.

(a) **Change-point analysis**

Change-point analysis is a statistical method used to determine where, if anywhere, in a sequence of observations, a change occurs in the value of some parameter (or parameters) of the underlying statistical model. Of the various methods available for change-point analysis, we have used the Bayesian approach of Smith (1975) which is further elaborated in Booth and Smith (1982). Cavanagh *et al.* (1988) demonstrate how the Bayesian approach can be applied to a series of phosphate concentrations taken at regular intervals along a transect across a site, in order to detect changes in the concentration corresponding to the site boundary as defined by phosphate content. In a further paper, Buck et al. (1988) describe how this method may be applied to a grid of values to detect changes in level. Initially, this method involves analysing each row and column of the data from the grid separately to allow detection of changes in level. Any part of the grid determined to have a high phosphate level by both row and column analyses, is deemed to have a high level of phosphate; otherwise the level is low. Thus the grid is divided into zones of high or low phosphate and so we have an initial estimate of the mean phosphate concentration both on and off site. In the second phase of the analysis, each row and column is reanalysed, but using the information gained about the means and standard deviations of the levels from the first calculations. Thus a new set of zones is defined. This process is repeated until the zones indicated remain identical from iteration to iteration.

(b) **Image Processing**

The image processing technique that we have used to analyse the data from Manor farm is that of Besag (1986). This method differs from the change-point analysis outined above in that it allows definition of more than two levels within the data. Although the method used for the change-point analysis could be extended to allow for more than two levels, the calculations involved would be immense and too time consuming, even on a mainframe computer; thus image processing is more practical. However, image processing requires that we specify the number of levels and, furthermore, that we provide good initial estimates of the mean and standard deviation of each level. Estimates of the number of levels and their parameters may be available from other comparable sites or, more likely, they will have to be obtained from the data under analysis. In practice the latter poses considerable problems as it may be viewed as determining the number of components (i.e. number of levels of phosphate concentrations) in a mixture (i.e. the whole data set) and then estimating the parameters of the components (i.e. the mean and standard deviation of each level). This problem has received considerable attention in the statistical literature (see Titterington et al 1985), but overall no best method is available. The method used could vary from application to application, but we outline how we have tackled the problem in the case study which follows. Then, having decided upon the number of levels and having obtained estimates of the mean and standard deviation of the phosphate concentration for each level, an initial estimate of the extent of the zones is obtained using maximum likelihood. The image so obtained is then processed to 'clean up' the picture thus defining zones of phosphate at various concentration levels. Processing continues until the zones remain identical from iteration to iteration.

Results

(a) SP170

In this instance a fairly simple question was asked of the data - distinguish two levels one of which we should define as 'on-site' and the other as 'off-site'. The method requires some initial input from the archaeologist in addition to the data. That is to say, some prior

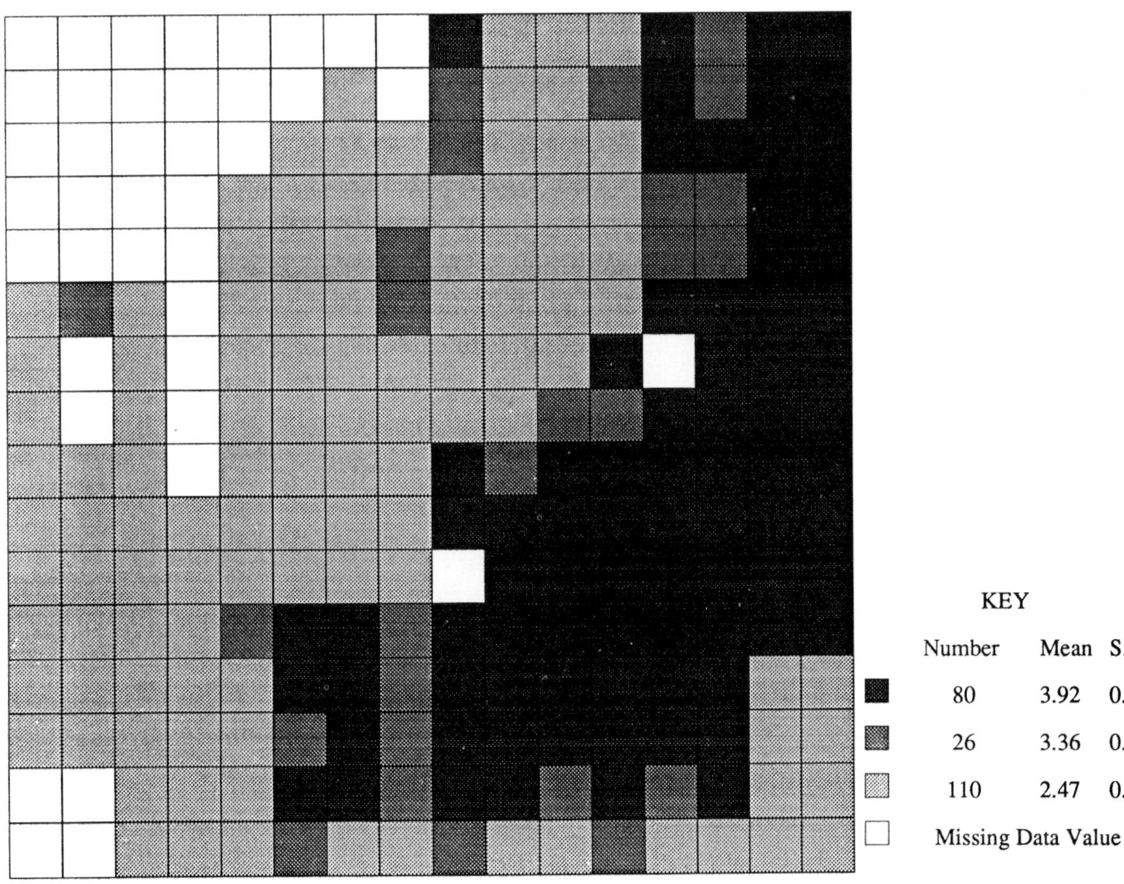

Figure 54 *Zones of low, medium and high phosphate concentration for site SP170 of the Laconia Survey.*

estimate of the probability of a change in level in any given row or column of data needs to be made; in our case a not unreasonable supposition would be that there is an equal probability of change or no change. In fact, after the process of iteration, even quite high probabilities of no-change still result in the same distribution of 'on' and 'off' -site phosphate. The results of the analysis are shown in Figure 54.

So we have a robust indication of what appears to be the 'phosphate' site. How is it to be interpreted? The site, as defined by phosphate, is larger than that indicated by pottery and tile. In particular there is a concentration along the eastern edge, which initially was interpreted as lying outside the site and caused by modern dumping beside the road, immediately to the east. In fact, investigation in 1988 indicated that contemporary archaeological remains are to be found on the other side of the road. It could be that in this case the phosphate has been a surer guide than sherd data or the archaeologist's initial estimate. It is hoped that geophysical tests will further aid the interpretation.

(b) **Manor Farm**

In the case of the Bronze Age cairn we are also faced with a site which is difficult to interpret. We indicated above that the samples were taken from an excavated surface. This was not strictly true in the sense that no old ground surface was found in the course of excavation. In fact the old land surface had become hopelessly mixed with the stones

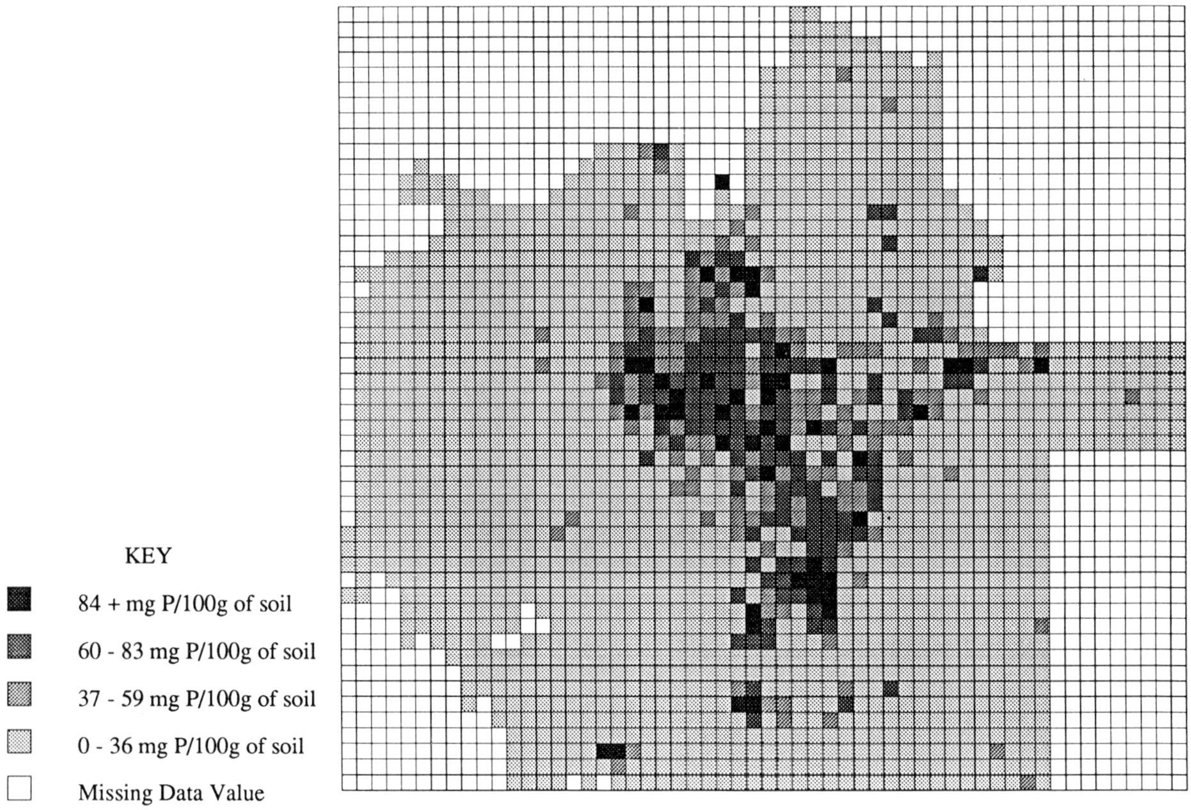

Figure 55 *Phosphate concentration zones determined by using the mean and standard deviation of the raw data from Manor Farm.*

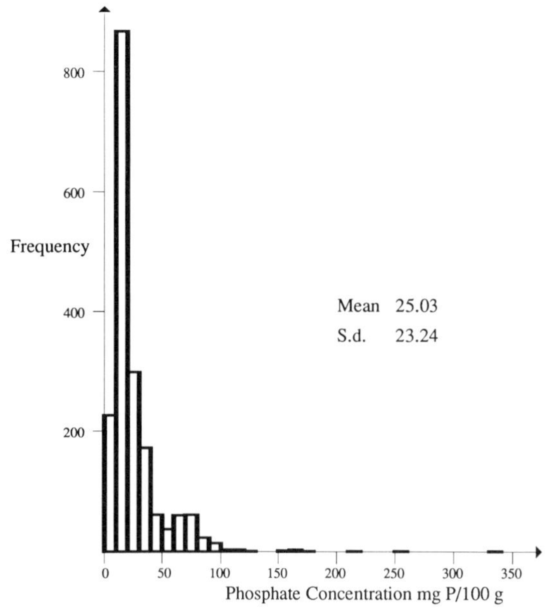

***Figure 56** Histogram of the soil phosphate concentration from Manor Farm.*

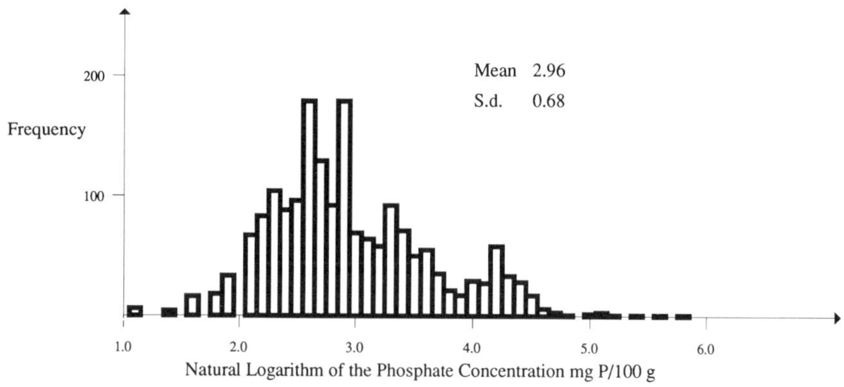

***Figure 57** Histogram of the natural logarithm of the soil phosphate concentration from Manor Farm.*

of the cairn through worm action and other post-depositional processes. Thus the samples were not taken, could not be taken, from a consistent archaeological surface. It was, nevertheless, clear from the beginning that enhanced phosphate was to be found within the cairn. An initial plot of the higher counts in the raw data indicated especially enhanced levels at the centre and some more isolated spots elsewhere. But the raw data are very difficult to interpret further. In particular, one of the motives which prompted the sampling, was to investigate whether any isolated depositions of bones had so totally decayed as to escape detection through excavation. Thus isolated clusters of higher phosphate would be important. Change-point analysis seems less appropriate as it is better suited for the detection of large contiguous blocks of high level, rather than small independent patches.

Image processing was the way forward, but rather more needs to be input by the archaeologist before the analysis can proceed. In the first place it is far from clear how many levels should be distinguished. In the published analysis of the raw data (Olivier 1987) four levels were chosen and the resulting grey scale diagram is given in Figure 55.

Figure 58 *Phosphate concentration zones for Manor Farm determined using image processing. The initial estimates of the mean and standard deviation of each level were (2.6,0.4), (3.5,0.4) and (4.3,0.4). The value of the smoothing parameter (ß) was 1.5.*

Inspection of the histogram of the raw data shows a slightly skewed distribution (see Figure 56). The histogram of the natural logarithm of the data (see Figure 57) is less skewed. It suggests that there are three levels of phosphate concentration and this forms the basis of our analysis. Secondly, initial estimates are required of the means and variances of the three levels - small differences in the choice of these means can lead to different results, some squares change their assignment to the high, medium or low levels, although the broad picture is reasonably robust. Again the mean levels chosen were estimated from the histogram. The initial estimates of the means were taken as 2.6, 3.5 and 4.3; the standard deviation of each level was initialised at 0.4. Immediately it is clear that image processing is a sensitive technique and a far greater discretion is left to the investigator.

But there is yet a third parameter which will govern the results. This is the value of the smoothing parameter ß, which, crudely put, decides the degree to which adjacent values will affect each other, the extent to which the assignment of any given square to a particular level is influenced by neighbouring squares. If ß is zero then adjacent squares affect each other not at all. As the value of ß increases so the picture effectively becomes smoother and smoother. In one sense it might be argued that in looking for isolated burials we ideally want a more fragmented and a less smooth picture. In fact change-point analysis was rejected as inappropriate because it made the image *too* smooth. Nevertheless, it was decided to proceed with a relatively high value of ß of 1.5, if only because this would make the results more dependable. The probabilities are, as it were, loaded slightly against the desired result making what does emerge that bit more reliable. The resulting image is given in Figure 58.

How then to interpret the results? In the first place there is a high concentration at the centre of the cairn - this was quite plain in the original raw data and is also indicated in the results of image processing. Although this central block of high phosphate includes the primary burial 1 (see Figure 17 of Olivier 1987, 182), in fact it does not markedly overlap with most of the identified burials, which tend to fall on the edges of the high phosphate area or within the medium phosphate sections. Thus the high phosphate block does not seem entirely consistent with the sought missed burials - perhaps there was some more intensive funerary use, such as disturbed or redeposited bone which has decayed (Olivier 1987, 181) or the exposure or cremation of bodies here. On the whole the majority of the observed burials coincided with higher levels of phosphate, though, for example, the burial SF61 in the SW corner does not. Image processing does, however, indicate some medium levels of phosphate to the west and north of the main concentration. It is not uncommon in the late bronze age for a circle of burials to be laid within a cairn and just such an interpretation of these zones has been suggested to us by the excavation director.

Conclusions

In conclusion, then, we believe that the statistical treatment of phosphate data offers a useful tool of interpretation to the field archaeologist. Undoubtedly there is an urgent need for fundamental research to be carried out into the soil chemistry of phosphorus compounds. In this light, much greater consultation with soil scientists is undoubtedly highly desirable. Regarding the statistical methods, we would stress the importance of two points in particular. First of all the selection of sampling interval and sampling strategy appropriate to the information sought. We have talked in terms of intervals varying from 0.5m to 10m, and in site prospection even larger intervals have been used (Sieveking *et al.*1973). We have also found gridded survey convenient, though change-point analysis does work in transects and could apply to radial transects. (This, of course, would be extremely sensitive to the correct location of the centre of the site). Secondly, the selection of an appropriate statistical method is crucial. As we have seen change-point analysis adopts a simple, perhaps over-simplistic, view of the spatial distribution, but has the advantage of robustness, or at least it requires fewer pre-suppositions on the part of the archaeologist. The image processing again operates from an intuitively simple principle, namely that adjacent data points are more likely to belong to the same phosphate concentration level. But its application requires a clear view on the part of the archaeologist of the number of levels involved, and relatively good initial estimates of the mean and standard deviations of the levels. This probably demands, at least for the immediate future, active interdisciplinary collaboration between the archaeologist, the chemist and the statistician in order to extract the most information from the data.

Acknowledgements

We are grateful to Adrian Olivier of the Cumbria and Lancashire Archaeological Unit for putting the Manor Farm data at our disposal and for his helpful suggestions in the interpretation of the image processing results. Thanks are also due to David Gurney of the Norfolk Archaeology Unit for his advice on the Manor Farm phosphate data. The chemical analyses of the Laconia Survey samples were carried out by Simon Hirst, Jill Symes and Andrew Fricker. We wish to express our thanks to the organisers of the conference and to those who made useful criticisms of our original presentation.

The soil phosphate analysis of site SP170 was carried out as part of the Laconia survey, a joint British - Dutch expedition, under the British School at Athens, directed by Dr H.W.Catling and led in the field by W.G.Cavanagh and J.H.Crouwel. Permission to carry out the survey was granted by the Greek Ministry of Culture. Thanks go to the staff of the Ministry and Dr T.Spyropoulos, Ephor of Antiquities for Laconia and Arcadia, for their kind permission and their assistance in the course of the survey. Financial support for the survey comes from the Managing Committee of the British School at Athens, the Society of Antiquaries, the British Academy, the University of Amsterdam Faculty of Arts, the Allard Pierson Foundation, the Amsterdam University Graduates Association and the Dutch Foundation for Pure Research (ZFO). The mathematics and computing that gave rise to this paper have been carried out under the auspices of a UK Science and Engineering Research Council project grant (reference: GB/E/30263).

References

Arrhenius O., 1938, 'Den Gotländska Åkerjordens Fosfathalt', *Sveriges Geologiska Undersokning*, **32**(2), 1-15.

Besag J., 1986, 'On the statistical analysis of dirty pictures', *J. Royal Stat. Soc.*, B, **48**, 3, 259-302.

Bintliff J.L., Davis B.T., Gaffney C. and Waters A., forthcoming, 'Trace metal accumulations in soil as indicators of ancient site occupation', *Environmental Geochemistry and Health,* **10**.

Booth N.B. and Smith A.F.M., 1982, 'A Bayesian approach to retrospective identification of change-points', *Journal of Econometrics* **19**, 1, 7-22.

Buck C.E., Cavanagh W.G. and Litton C.D., 1988, 'The spatial analysis of site phosphate data' In *Computer Applications in Archaeology 1988*, S.P.Q. Rahtz (ed.), BAR International Series **446**(i), 151-160.

Cavanagh W.G., Hirst S. and Litton C.D., 1988, 'Soil phosphate, site boundaries and change-point analysis', *Journal of Field Archaeology*, **15**, 1, 67-83.

Conway J.S., 1963, 'An investigation of soil phosphorus distribution within occupation deposits from a Romano-British hut group', *Journal of the Archaeological Society*, **10**, 117-128.

Craddock P.T., Gurney D., Pryor F. and Hughes M.J., 1985, 'The application of phosphate analysis to the location and interpretation of archaeological sites', *Archaeological Journal* **142**, 361-376.

Dietz G.F., 1957, 'Phosphorus accumulation in soil of an Indian habitation site', *American Antiquity* **22**(4), 405-409.

Fieller N.R.J. and Flenley E.C., 1988, 'Statistical analysis of particle sizes and sediments', In *Computer and Quantitative Methods in Archaeology 1987*, C.L.N. Ruggles and S.P.Q. Rahtz (eds.), BAR International Series **393**, 79-94.

Hamond F.W., (1983), 'Phosphate analysis of archaeological sediments', In *Landscape Archaeology in Ireland*, T. Reeves-Smith and F.W. Hamond (eds.), BAR British Series **116**, 47-80.

Keeley H.C.M., 1981, 'Recent work using soil phosphorus analysis in archaeological prospection', *Revue d'Archéométrie* **5**, 59-95.

Olivier A.C.H., 1987, 'Excavation of a Bronze Age funerary cairn at Manor Farm, near Borwick, North Lancashire', *Proceedings of the Prehistoric Society* **53**, 129-186.

Sieveking G. de G., Longworth I.H., Hingles M.J., Clark A.J. and Millet M., 1973, 'A new survey of Grimes Graves, Norfolk', *Proceedings of the Prehistoric Society* **39**, 182-218.

Smith A.F.M., 1975, 'A Bayesian approach to inference about a change-point in a sequence of random variables', *Biometrika* **62**, 2, 407-16.

Titterington D.M., Smith A.F.M. and Makov U.E., 1985, *Statistical Analysis of Finite Mixture Distribution.*, Wiley, Chichester.

Widgren M., 1983, 'Settlement and farming systems in the early Iron Age', *Stockholm Studies in Human Geography* **3**, Almquist and Wiksell International, Stockholm.

Multielectrode Resistivity Tomography for Imaging Archaeology

Mark Noel

Department of Archaeology, University of Durham.

Abstract

Conventional geophysical surveys produce 2-D area maps of archaeological features in terms of a contrast in some physical property. The absence of depth information hinders the interpretation at sites with deep or complex stratification or where burial depth alters significantly. A new approach to electrical resistivity prospecting uses a multielectrode array to provide depth as well as area information. Potentials are recorded along the array when currents are driven between successive pairs of electrodes. With N electrodes there exist N(N-3)/2 possible independent measurements of apparent resistivity which are used to reconstruct a vertical 'image section' by backprojection. Theoretical, numerical and field studies are leading to the design for a new instrument which will help resolve archaeological features in three dimensions.

Introduction

All geophysical prospecting methods rely on detecting a difference in physical properties between buried features and surrounding media. The earliest geophysical attempts to locate archaeological remains were based on measuring variations in magnetic field strength or ground electrical resistance. Although alternative methods have since been evaluated, these two techniques remain the mainstay of modern geophysical surveys in archaeology (Aitken 1974).

The electrical conductivity of geological materials is mainly a function of water content since most charge is transferred by ionic conduction. Electrical resistivity prospection measures changes in ground resistivity expressed in terms of the ratio between the potential measured at the surface and the applied current. To avoid problems associated with contact resistance, polarization effects and earth currents, it is usual to employ a four probe system and an alternating current supply. With this arrangement, a number of probe configurations are possible but the Wenner array was the first to be adopted for archaeology after extensive use in geology.

Since resistivity surveying requires the repeated insertion and removal of probes, the technique is intrinsically slower than magnetic methods. However, after the twin electrode array was introduced around 1970 the method became more flexible and anomalies were also simpler to interpret in terms of archaeological structures. At the same time, field apparatus has improved through advances in electronics and datalogging, while computer programs now enable results to be presented in a variety of formats with features enhanced by image processing.

Despite these recent developments in instrumentation, it is important to recognise the limitations inherent in conventional area survey techniques and their effect on archaeological interpretation:

1 The resistivity survey provides only a 2-D map with no representation of the vertical stratigraphy. This disadvantage becomes particularly important when attempting to decipher complex urban archaeological deposits.

2 The probe spacing is chosen to give a compromise between spatial resolution and 'depth of investigation'. Hence, the geophysical appearance of archaeological features can seldom be optimised over the whole survey area.

These problems could be overcome if the resistivity survey method was modified to provide both vertical and horizontal stratigraphic information. A partial solution might be to carry out a number of independent surveys, changing the probe spacing and thus the depth of investigation. The result would then be a set of resistivity maps biased towards different levels from which some stratigraphic information could be extracted. However, because this technique would be expensive in terms of survey time it has not been widely applied.

For geological purposes, the traditional approach to stratigraphic modelling has been to use resistivity measurements on an expanding 4-electrode array. The idea is simply to increase the current penetration and so bias the apparent resistivity values towards strata at greater depth. If the measurements are repeated a number of times, with the centre of the array displaced, the result will be a 'pseudo earth section' providing a crude picture of resistivity structure in a vertical plane (Griffiths & King 1981). A number of commercial systems are now available which use microprocessors and electronic switching to automate the collection of resistivity data and compute these pseudo sections (e.g. Griffiths & Turnbull 1985). Finite element modelling techniques have also been developed to deduce a more accurate geological structure from expanding array data (Dey & Morrison 1979). However, the technology has yet to be systematically applied in archaeology and is unlikely to yield resistivity sections with adequate resolution. This paper describes a new approach to resistivity prospection in which geophysical sections through the archaeology are reconstructed tomographically, using measurements on a multielectrode array.

Electrical Resistivity Tomography

This project aims to develop a geophysical system capable of producing resistivity sections at high resolution together with area maps at different depths. Here an important stimulus has been recent research in medical physics which has led to the new imaging technique of Applied Potential Tomography, or APT (Brown 1986). In this method, the distribution of electrical conductivity in an anatomical section is computed from voltages developed on an encircling electrode array in response to currents passed through consecutive pairs of electrodes. The first practical scanner uses 16 skin electrodes to produce 104 independent voltage measurements from which the resistivity cross-section is computed in real-time using a fast backprojection algorithm (Brown & Seager 1987, Barber & Seager 1987).

Preliminary work by Powell, Barber and Freeston (1987) suggests that the principles underlying medical APT can be adapted to a linear array of electrodes for archaeological surveying. With N electrodes, there exist N(N-3)/2 possible independent measurements of apparent resistivity, which here we refer to as the Data Set. All these measurements are gathered in order to permit an optimum reconstruction of the resistivity section. This compares to the classical Wenner pseudo section method where the data set is incomplete. It can be shown that the data set can be specified in (N-1) possible ways depending on the separation of the current drive electrodes. Figure 59 shows two steps in the measurement sequence for data set 1, (inter-electrode spacing = 1).

The potentials recorded at each step are used to adjust the resistivities of those cells (pixels) which lie between equipotentials extending from the measurement electrodes (Figure 1; Barber & Seager 1987). Projection of measured potentials for the entire data set thus completes the inversion which can be further refined by iteration. This method is attractive for its stability, simplicity and speed of computation. The term 'tomography' is quite appropriate here since the resistivity section is being reconstructed by projecting boundary data (Gordon *et al.* 1975). Computed tomography is also applied to seismic X-ray, particle and acoustic imaging (Sochurek & Miller 1987).

The advantages that electrical resistivity tomography (ERT) would have over the conventional pseudo earth section method are:

1 By obtaining all possible independent potential measurements on the array (rather than just Wenner subsets) a more complete electrical description of the subsurface must be obtained.

2 A measurement sequence that involves the complete data set generates a more isotopic distribution of current vectors at pixels in the subsurface than for an expanding Wenner array. This is fully explained in Figure 60. The shape of archaeological objects will therefore be reconstructed at higher resolution and with much less distortion.

3 Any desired inter-electrode potential can be calculated from the complete data set by arithmetic combination and used as the basis for a conventional area resistivity map. Tomographic section data could also be merged to give 3-D block representations or area maps at different depths.

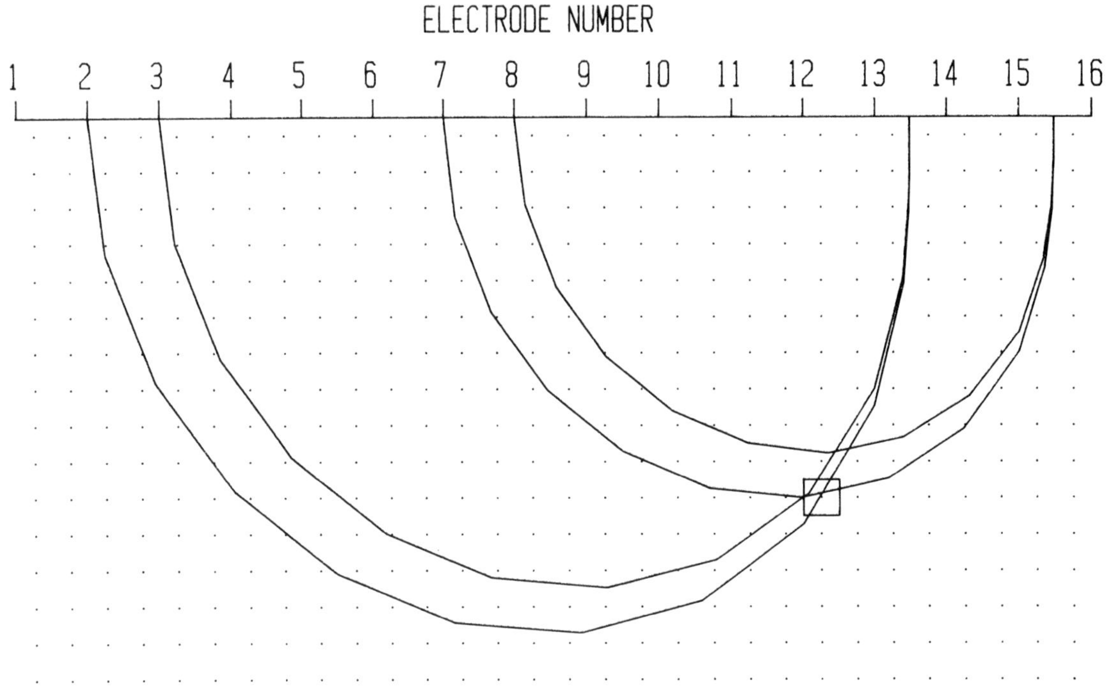

Figure 59 *The method of backprojection used in generating resistivity sections. Two stages in the image reconstruction are shown for a single pixel (small square) beneath a 16 electrode array. The potential measured between electrodes 2 & 3 when currents are passed between 13 & 14 is compared to the potential expected for a homogeneous region. The pixel density is adjusted according to this ratio (with weighting factors) and the operation repeated as a sum for the measurement: 7, 8, 15 & 16, and all other projections which intersect the pixel.*

In research towards a medical system, Powell et al (1987) succeeded in reconstructing resistivity images of objects suspended in a saline tank using a 16 electrode linear array and simple backprojection. A resistivity prospection system using electrodes inserted in boreholes beneath the site is under development by the Oyo Corporation of Japan (Sakayama and Shima 1988, Shima & Sakayama 1988). Sections are computed by a simplified backprojection that assumes linear equipotentials but it is unlikely that such an invasive technique would prove acceptable for archaeological surveying.

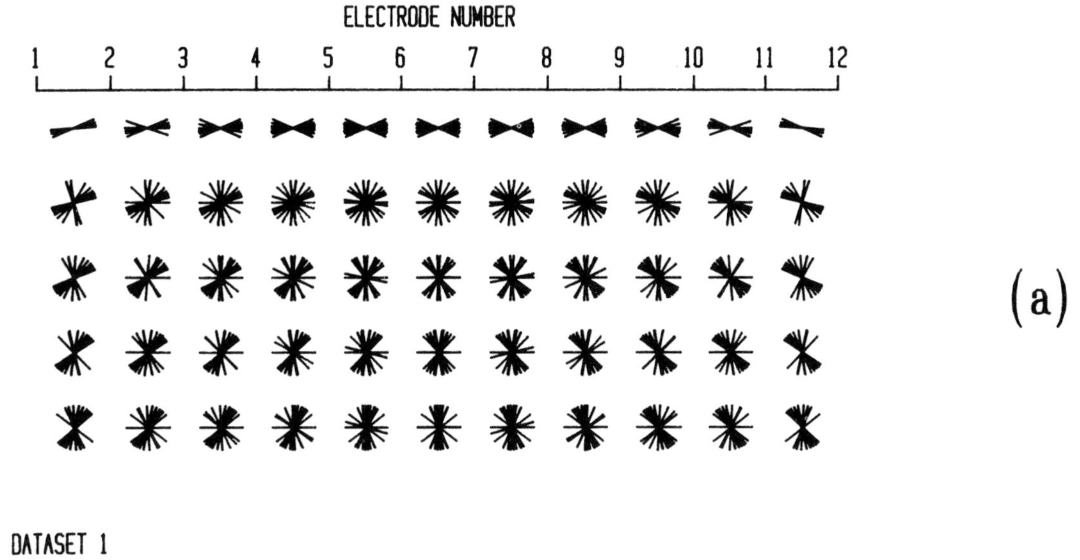

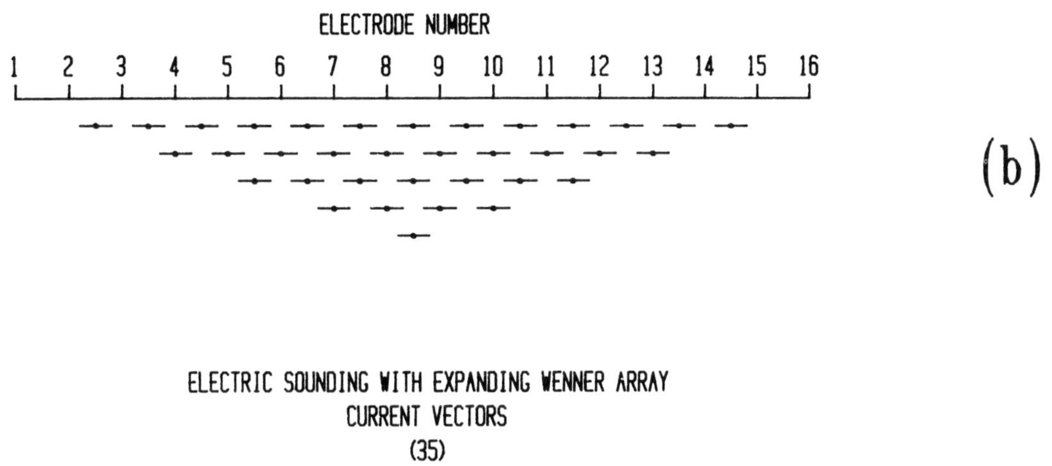

Figure 60 *(a) The directions of current vectors associated with all backprojections which pass through individual pixels. The situation is shown for an array of only 12 electrodes for the sake of simplicity. (b) The pattern of current vectors induced by an expanding Wenner array when used for resistivity sounding. Each vector has been plotted at the depth conventionally assigned to the apparent resistivity, i.e. at a depth of half the interelectrode separation and beneath the midpoint of the Wenner array.*

Evaluation of Method

Numerical Simulation

A computer model has been used to simulate the potential distribution measured along a linear electrode array in response to simple buried structures. Geometries examined include a vertical wall or boundary between regions of differing conductivity; two horizontal layers and; a small buried sphere. Methods for calculating the boundary potentials have been described in detail elsewhere (e.g. Keller & Frischknecht 1966). Measured Data Sets were then backprojected to reconstruct resistivity sections which could be compared to the input models. Results for a 16 element array are shown in Figure 61. The numerical simulations confirm that this simple method of inversion is able to provide a reasonable estimate of the resistivity section, at least for features that are shallow compared to the length of the array. Image 'streaking' and distortion near the boundaries of the image section are due to current anisotropy (Figure 60a) and a rapid fall in sensitivity to resistivity changes within this region. Improvements to the inversion method along with image processing techniques would help to reduce these artefacts.

Field Trials

A preliminary trial of electrical resistivity tomography surveying has been carried out at Verulamium where excavation provides prior evidence for the location and depth of Roman structures (Collingwood & Richmond 1969). The experimental system comprised 16 electrodes of the Martin-Clark design and an Abem Terrameter SAS300. Connections between the resistivity meter and the electrodes were changed by hand. Each Data Set (104 readings) took approximately one hour to measure and record.

Two separate experiments were carried out in order to evaluate the overall concept and the performance of the image reconstruction algorithm (Figure 62). A test area was selected near the hypocaust building, (Insula IV) and in both cases the interelectrode spacing was 75 cm.

1 Two sets of measurements were first obtained with the centre of the array crossing a single wall at right angles and with a sideways displacement of 2m (Figure 62a). Here the objective was to test whether this resistive feature would appear at a consistent position in the reconstructed images. These results are shown in Figure 63a.

2 The array was then positioned across two adjacent, parallel walls and sets of measurements obtained as the array traversed these features at right angles (Figure 62b). Forward movement of the array was achieved by 'leapfrogging' the rear electrode in the conventional manner. In this test, resistive features appearing in the image section should displace with movement of the array. Results are shown in Figure 63b.

The two resistivity sections produced in the first experiment reveal a fixed, high resistivity feature near the mid-point of the array with a position and scale consistent with the known hypocaust wall. In Figure 63b two high resistivity features are seen to move through the resistivity section as predicted from the experimental geometry. These results confirm that the ERT method is capable of producing resistivity cross-sections through archaeological structures.

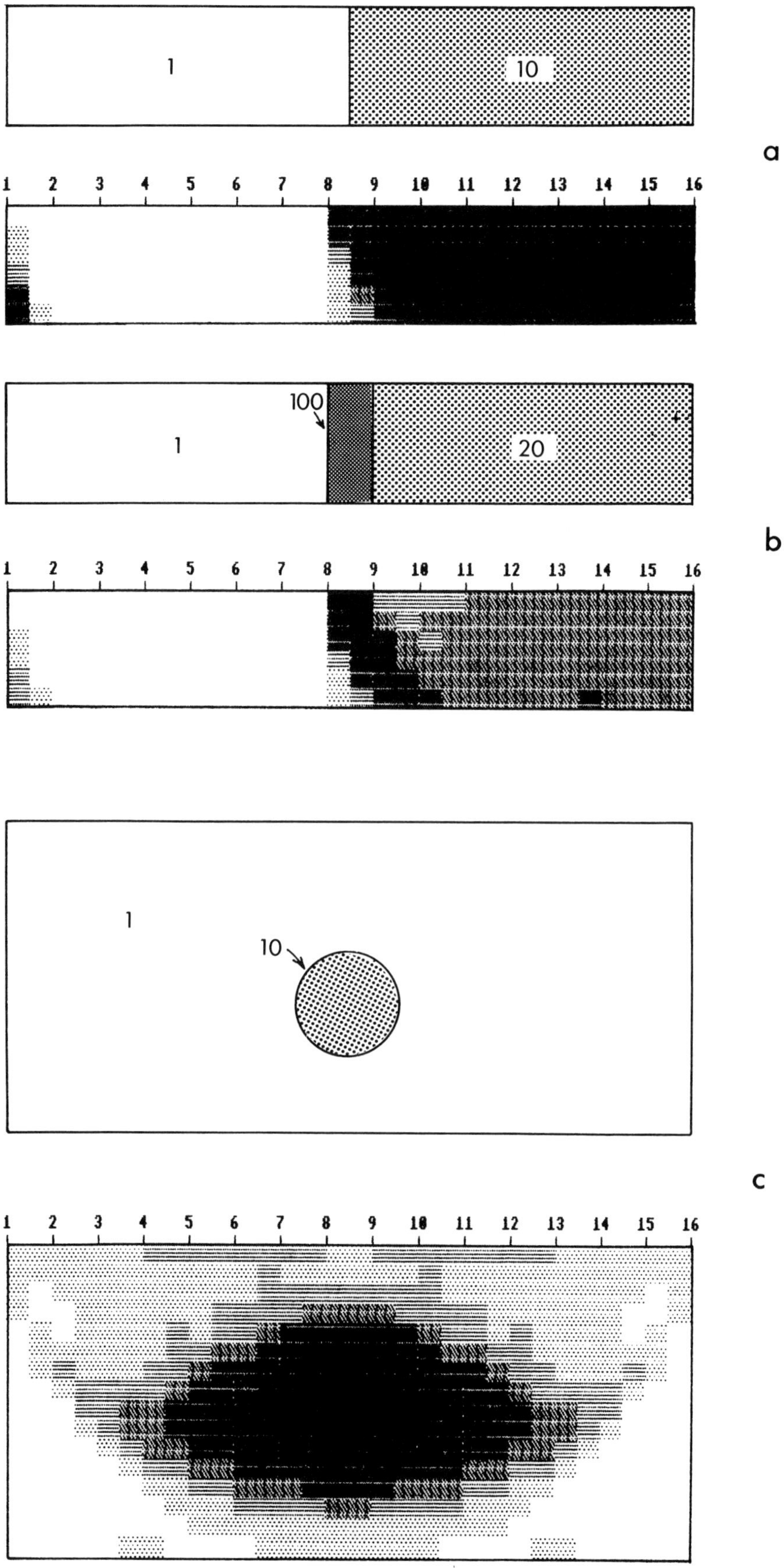

***Figure 61** Computer reconstruction of resistivity sections through model structures based on a simple backprojection algorithm: (a) A vertical boundary between two regions of contrasting resistivity. (b) A wall of high resistivity flanked by two, differing, lower resistivities. (c) A small, more highly resistive sphere. The numbers give the relative resistivities (arbitrary units).*

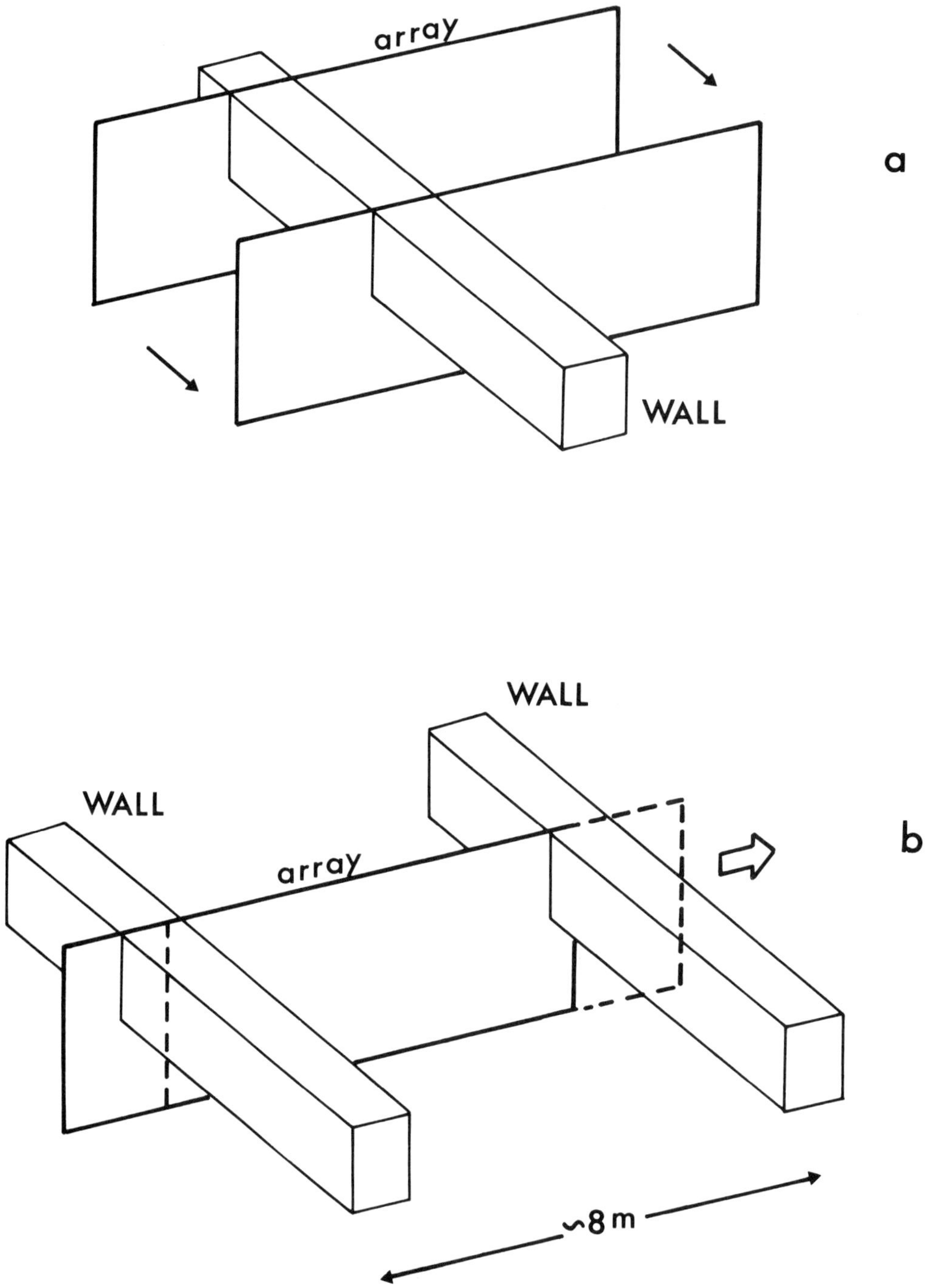

Figure 62 *Geometries of the two field tests carried out near the hypocaust building at Verulamium (neither to scale). The interelectrode spacing was 75 cm. (a) Two data sets were collected over a wall, approximately 75cm wide, with the 16 electrode array displaced sideways 2m. Archaeological evidence suggested that the depth to the top of the wall was about 15cm. The end of the array was stationed 1-2m from the midpoint of the wall. (b) A series of data sets were collected with the same array traversing two shallow walls at right angles. Dimensions and depths of these walls were similar to the previous example and were separated by ~ 8m. The traverse experiment started with the end of the array positioned 1-2m from the midpoint of the leftmost wall.*

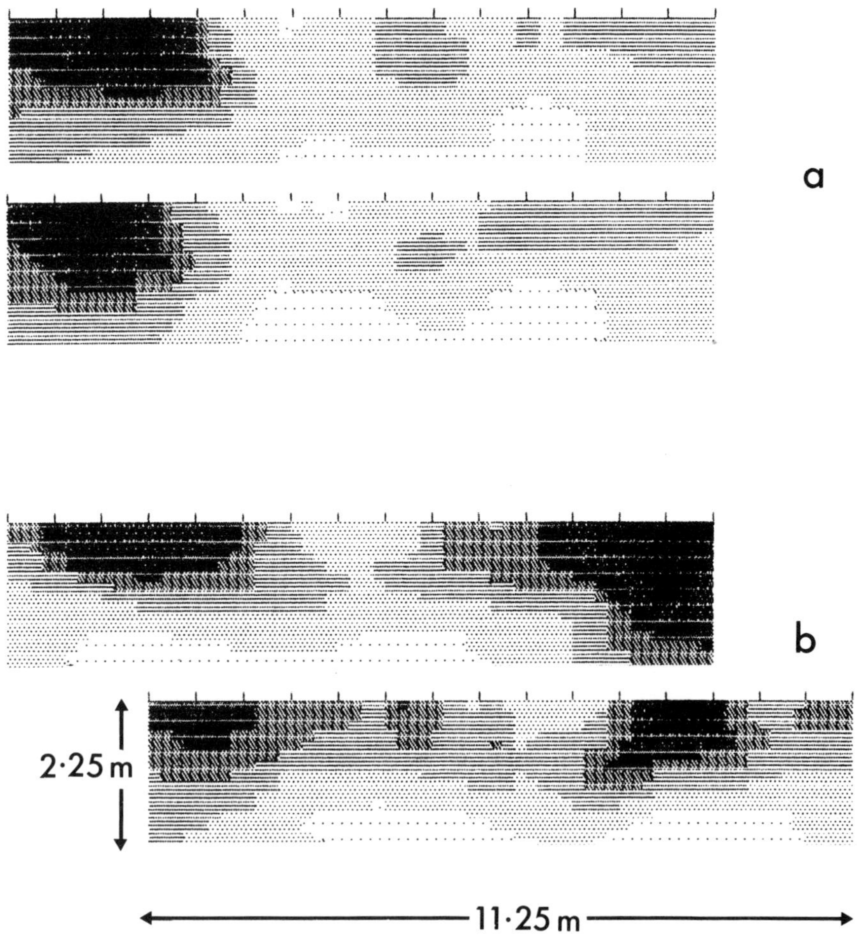

Figure 63 Results of the two field trials at Verulamium. (a) The two resistivity sections obtained from the experiment shown in Fig 62a. The vertical pips indicate the electrode positions and the image dimensions are as shown. Dark regions correspond to zones of higher resistivity (the wall). (b) Two resistivity sections obtained with the array displaced to the right and showing relative movement of two walls (dark areas). Experimental geometry as in Figure 62b.

Discussion

Results from the numerical experiments and preliminary field trials encourage further development of electrical resistivity tomography as a new prospection tool. In archaeological and medical applications simple backprojection provides a rapid estimate of the impedance cross-section from the measured potentials. However, other numerical techniques may provide a more accurate inversion and eliminate some of the artefacts apparent in the images shown here (Yorkey & Webster 1987). These include the Newton-Raphson algorithm (Yorkey 1986) or sensitivity and finite element methods (Murai & Kagawa 1985, Dey & Morrison 1979).

Undoubtedly, the principal obstacle to implementing ERT surveying with conventional equipment is the time required to gather such a large data set. This difficulty has already been overcome in the medical application where specialised multiplexing circuitry linked to 16 electrodes can record anatomical APT data in as little as 79 milliseconds (Brown and Seager 1987). Such high speeds would be hard to achieve with an archaeological system because low A.C. frequencies and signal averaging are necessary to reduce errors arising from earth currents, etc. Nevertheless, with a specialised resistivity meter incorporating a data logger, it seems certain the ERT measurements could be gathered at a sufficient rate to make this a powerful and attractive technique for routine survey.

Acknowledgements

I am particularly grateful to I. Noel for playing the part of multiplexer during the field trials. R. Nesbitt and C. Saunders of the Verulamium Museum gave permission and advice for the field survey. Data processing was accomplished using a portable computer kindly loaned by Sharp UK. D. McCann and G. Ripin of the British Geological Survey generously provided the Abem Terrameter used in this research.

References

Aitken M.J., 1974, *Physics and Archaeology*, Oxford University Press.

Barber D.C. and Seager A.D., 1987, 'Fast reconstruction of resistance images', *Clinical Phys. Physiol. meas.* **8**, 47-54.

Brown B.H., 1986, 'Applied potential tomography', *Phys. Bull.* **37**, 109-112.

Brown B.H. and Seager A.D., 1987, 'The Sheffield data collection system', *Clinical Phys. Physiol. meas.* **8**, 91-98.

Collingwood R.G. and Richmond I., 1969, *The Archaeology of Roman Britain*, Methuen, London.

Dey A. and Morrison H.F., 1979, 'Resistivity modelling for arbitrarily shaped two dimensional structures', *Geophys. Prosp.* **27**, 106-136.

Gordon R., Herman G.T. and Johnson S.A., 1975, 'Image reconstruction from projections', *Sci. Amer.* **223**, 56-68.

Griffiths D.H. and King R.F., 1981, *Applied Geophysics for Geologists and Engineers*, p.101 Pergamon, Oxford.

Griffiths D.H. and Turnbull J., 1985, 'A multi-electrode array for resistivity surveying', *First Break* **3**, 16-20.

Keller G.V. and Frischrnecht F.C., 1966, *Electrical Methods in Geophysical Prospecting*, Pergamon, Oxford.

Muria T. and Kagawa Y., 1985, 'Electrical impedance computed tomography based on a finite-element model', *IEEE Trans. Biomed. Eng.*. **BME-32**, 177-184.

Powell H.M., Barber D.C. and Freeston I.L., 1987, 'Impedance imaging using linear electrode arrays', *Clinical Phys. Physiol. meas.* **8**, 109-118.

Sakayama T. and Shima H., 1988, 'High resolution two dimensional resistivity inversion technique using alpha centers', *Electromagnetics* **1.3**, 47-49.

Shima H. and Sakayama T., 1988, 'Resistivity tomography: An approach to 2-D resistivity inverse problems', *Electromagnetics* **1.4**, 59-61.

Sochurek H. and Miller P., 1987, 'Medicines' new vision', *National Geographic* **171**, 2-41.

Yorkey T.J., 1986, *Comparing reconstruction methods for electrical impedance tomography*, Ph.D. Thesis, University of Wisconsin, Madison, WI.

Yorkey T.J. and Webster J.G., 1987, 'A comparison of impedance tomographic reconstruction algorithms', *Clinical Phys. Physiol. meas.* **8**, 55-62.

Recent Developments in Thermal Archaeological Prospection

T.J. Bellerby*, M. Noel, K. Brannigan*****

**Department of Geology, University of Sheffield (Now at Remote Sensing Unit, Department of Geography, University of Bristol);*
***Department of Geology, University of Sheffield (Now at Department of Archaeology, University of Durham);*
****Department of Archaeology and Prehistory, University of Sheffield.*

Abstract

Aerial infrared thermography is becoming an established means of archaeological prospection. However, such airborne techniques do suffer from high cost and interference caused by differences in the properties of the ground surface. An alternative thermal prospection technique is suggested, using direct contact temperature measurements with a simple probe at a depth of 20cm. The physical theory, field techniques and equipment developed for the technique are described, and the results from successful field trials at Verulamium presented.

Introduction

The flow of heat in the shallow (<20m) part of the ground is dominated by the effects of the periodic diurnal and annual changes in ground surface heat balance. Geological bodies interact with deeper geothermal fluxes; the temperature anomolies associated with these structures being locatable either through measurements taken below 20cm or by removing the effects of surface variation from more shallow temperature readings (eg. Poley & Van Steveninck; 1970, 1971, Kappelmayer 1957). Most archaeological features lie within the top 20m layer and thus the associated thermal anomolies are both created by and superimposed upon the surface driven heat flow pattern.

Benner and Brodkey (1984) have compared the results of a numerical model for the interaction between subsurface features and the diurnal variation with results from an infrared sensor measuring surface temperatures over buried test objects. For very shallow objects (<6cm) they found, and interpreted, detectable anomalies. Perriset and Tabbagh (1981) have modelled the effects of deeper features upon longer duration changes in heat flux and conducted tests of an airborne infrared camera which successfully located Neolithic ditches, pits and field boundaries.

The detection of buried archaeological features by aerial thermography is now becoming an established procedure. The interaction of natural heat fluxes with features possessing contrasting thermal properties will create anomolies in the surface temperature which are located using the airborne infrared camera. However, airborne thermographic systems are relatively expensive and readings are affected by variations in surface emmisivity, the composition and state of the atmosphere. These disadvantages would not apply to a thermal technique based on direct measurement of ground temperatures at 20cm depth using a simple thermistor probe.

The implementation of a direct contact technique presents a number of new problems. The existing theory covering the generation of anomalies in surface temperature must be extended to cover temperatures at 20cm depth. An extended survey will encounter the continuing diurnal variation which must be corrected for, together with the effects of probe frictional heating. A satisfactory solution has been found to each of the above problems and a practicable field technique, hardware and data processing methods, have been developed and then tested at Verulamium.

Theory

As noted above, the heat flux around archaeological features is predominately generated by thermal interactions at the ground surface. Since variations in the surface heat balance possess strong periodic diurnal and annual components, they may be represented by a periodic function, written in terms of a Fourier series (Van Wijk, 1963). The linearity of the heat diffusion equation allows the contributions due to each harmonic to be studied separately.

A harmonic of the surface temperature variation will create a temperature 'wave' which exponentially decays in magnitude with increasing depth (Figure 64). For harmonics of a greater period this decay is less pronounced and the wave penetrates a greater distance into the ground. In typical conditions, the diurnal harmonic is reduced to ~1% of its surface magnitude at a depth of 60cm while the corresponding depth for the annual harmonic is about 21m.

The presence on the soil of a body with differing thermal properties disturbs this sample pattern. Towards the centre of a wide body the heat flow may be regarded as being the same as that through an infinitely wide, three layered structure, a problem which may be solved analytically in one dimension (Van Wijk, 1963).

Comparing the results for this layered structure with those for homogenous ground allows an approximate value for the temperature anomaly to be obtained at a depth over the body. Figure 65 shows the estimated peak magnitudes for temperature anomaly components plotted against the period of the associated harmonics.

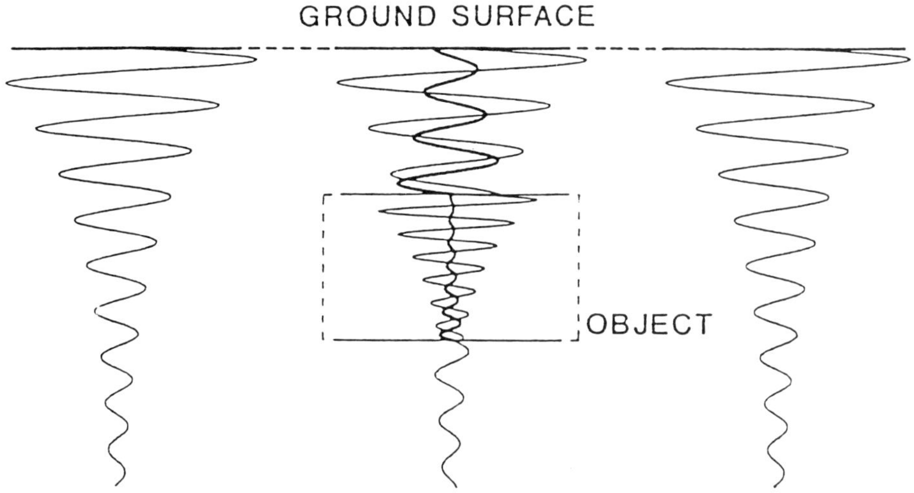

__Figure 64__ A schematic representation of periodic heat flow interacting with a buried object of rectangular cross-section. The thin lines represent the decaying downgoing temperature wave; the thick lines the reflected temperature wave.

A more detailed examination of the shape and magnitude of ground temperature anomalies has been carried out using a two dimensional finite-element technique. Figure 66 shows the results for simulated traverses across an elongated body of rectangular cross section such as a wall or ditch for a selection of body depth to width ratios. Each anomaly component varies periodically with the same period as the associated harmonic but with a phase difference depending upon feature depth, harmonic period and soil properties. The one year component is essentially in phase with the annual temperature cycle, peaking in midwinter and midsummer.

Carson (1963) has applied Fourier techniques to records of ground temperature data. The primary component of the annual wave was shown to comprise at least 95% of the total variation. It may be deduced that the 1y harmonic will be the dominant influence in generating any 20cm ground temperature anomaly associated with objects deeper than ~40cm. In midsummer and midwinter the sign of the measured anomaly is wholly determined by the annual component which provides a larger contribution than that of the other components combined. This fact allows the sign of the measured anomaly to be related to the contrast in thermal properties between the soil and the feature at these times of the year. During other seasons and for shallower objects such an interpretation would be more complex and require much more information about the past variation in the surface heat flux or temperature with time.

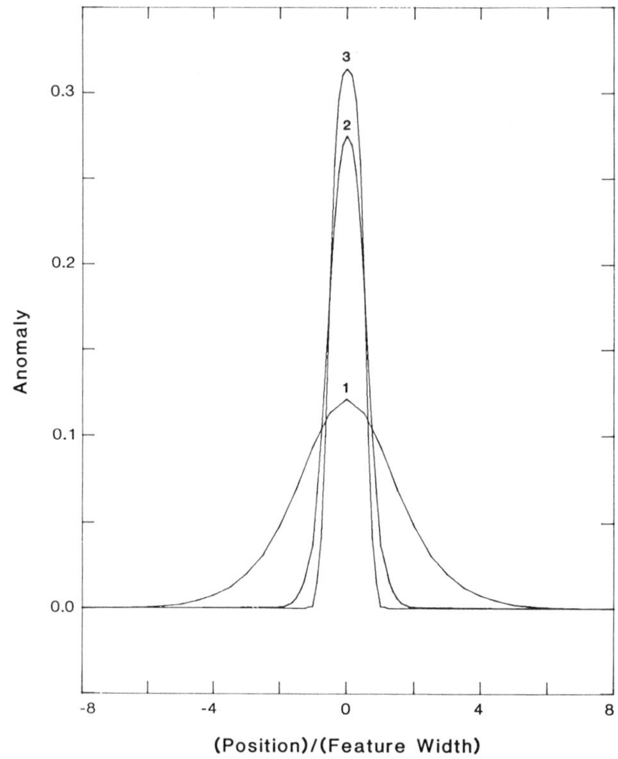

Figure 65 *Estimates of peak soil temperature anomalies associated with a body of water saturated sandy soil buried in peat. Curves 1,2, ..6 correspond to feature depths of 20cm, 30cm, ..70cm. Anomaly magnitudes are given in °C for a 10°C peak to peak surface variation.*

Field Equipment and Data Processing

Three main problems must be overcome in developing this thermal prospection technique. A thermometer probe must be developed to take temperature readings to the required accuracy. If the probe were allowed to reach effective equilibrium after insertion the rate of data collection would be impractical. Also the required 20cm measurement depth is within the effects of the diurnal temperature variation which will add an unwanted, secular component to the survey data.

The accuracy of a shallow-ground temperature probe is limited by the extent to which the measurement depth may be controlled. For a 20cm measurement depth, positioning errors of +/- 1cm may occur, giving rise to temperature errors of +/- 0.1-0.2°C, while a typical ground temperature anomaly will be in the region of 0.3-1.0°C. In order not to further degrade this signal to noise ratio, a measurement system with an accuracy much better than this is required (Noel and Bellerby, 1990). A block diagram of a thermometer probe and data logging system developed to these specifications is shown in Figure 67. The probe itself contains a precision thermistor housed in a 4mm brass and stainless steel nose cone extending from a tufnol rod and is inserted into a hole made previously by a more mechanically robust steel probe. The thermistor comprises one of the timing components in an astable oscillator circuit feeding the frequency input of a Grant Squirrel 1201 Data Logger. The logger takes frequency readings over a period of one second, allowing a considerable averaging of any fluctuations in the oscillator circuit or thermistor. Long term drifts in circuit characteristics are corrected for by comparing the frequencies produced by a precision resistance, electronically switched in place of the thermistor.

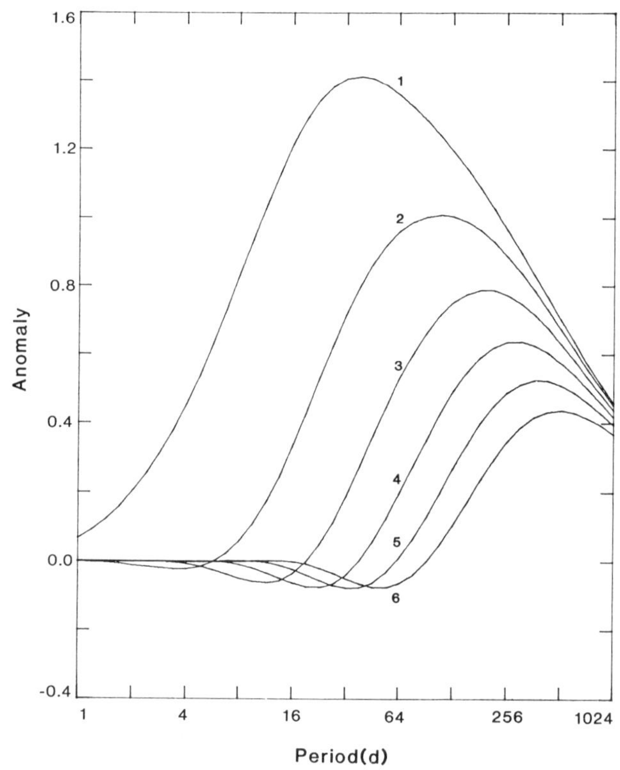

Figure 66 *Example anomaly shapes for a rectangular object at a depth of 1m, 10 days after a step change of 10°C in effective surface temperature. Curves 1, 2 and 3 correspond to object width/depth ratios 0.5, 2 and 4 respectively. Anomaly values are given in °C and will scale in proportion to the magnitude of the original step change.*

Although designed to minimise the thermal time constant, the probe takes about five minutes to achieve effective thermal equilibrium with the ground. It is not reasonable to wait for such a period in order to be able to make a temperature measurement. To overcome this problem a novel technique adapted from methods used in marine heat flow work (Hyndman et al. 1979) has been adopted. The true ground temperature is deduced from a series of readings taken after the insertion of the probe but before effective equilibrium is reached. This deduction comprises the least squares fitting of the reading sequence to an empirical expression for the equilibration. Figure 68 shows four such empirical curves fitted to data points lying between 10 and 40 seconds after insertion. The curves fit the complete data sets to within 0.04°C and will yield the ground temperature to a similar accuracy. The required measurement time of 40 seconds is still relatively long but this problem may be alleviated through the simultaneous use of more than one probe linked to a single data logger.

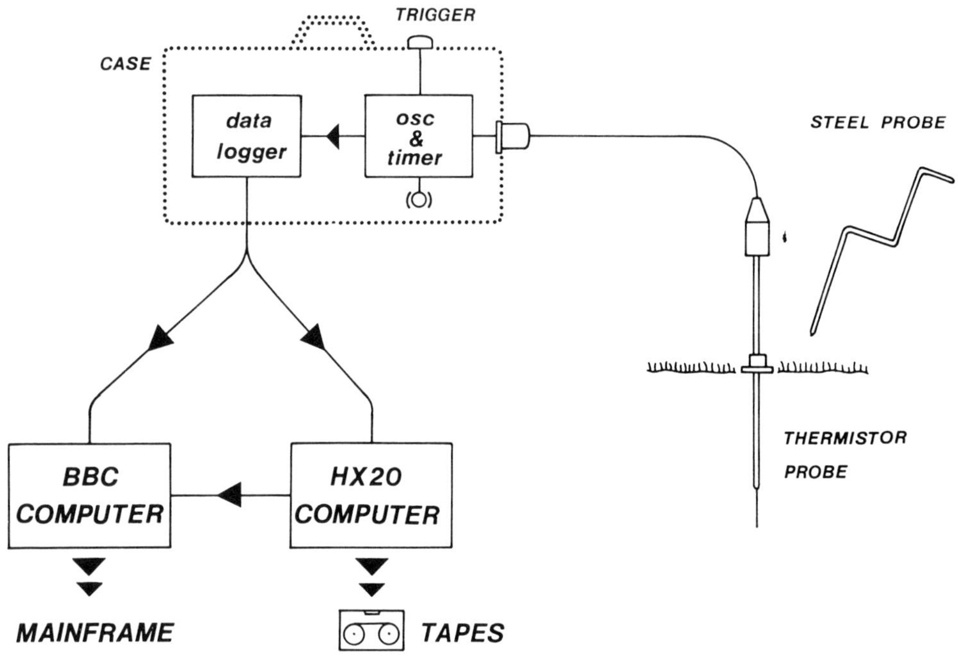

Figure 67 *A schematic representation of the main components in the thermal prospection system.*

The derived ground temperature readings remain superimposed upon the diurnal temperature variation which must be determined to a high accuracy in order that it's removal will not degrade the data. A sequence of readings taken at a single point (or base station) will give the form of the variation but measurement repeatability problems unacceptably degrade its quality. To improve this situation, use is made of the fact that the shorter-period harmonics of the diurnal variation do not penetrate to the measurement depth. Thus the removal of any higher harmonics present in the base station record will yield an improved representation for the diurnal variation.

Field Results

Field trials have been carried out at Verulamium (Roman St. Albans). This site lies under grass playing fields and thus has the advantage of simple conditions at the ground surface.

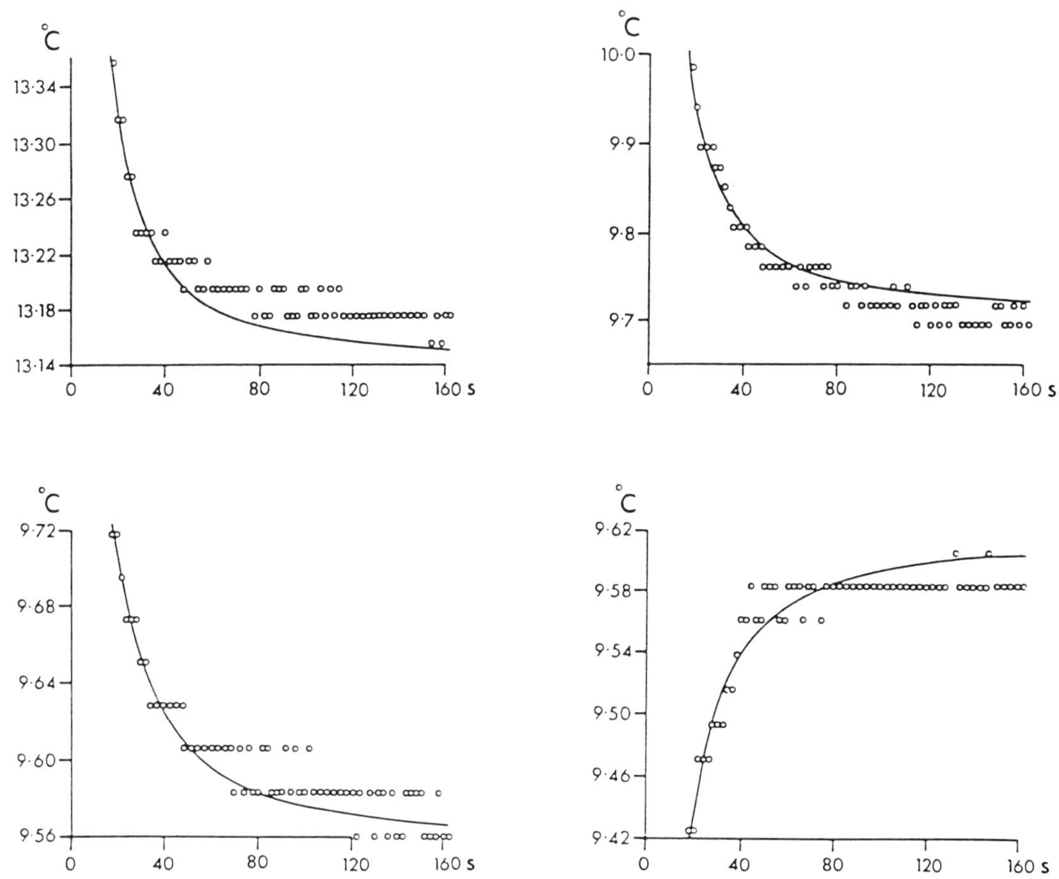

Figure 68 *Four example comparisons of actual and fitted equilibration curves for the thermistor probe. The solid curves were derived by fitting an empirical functional form to the first 40s of data, ignoring the first 14s.*

A comparison between thermal survey over one corner of the Insula III2 building and a corresponding resistivity survey is shown in Figure 69. Although previous excavation of this site has disturbed it too greatly for either technique to give a clear picture of the building, it can be seen that there is a definite correlation between the two sets of results. The curved feature on the bottom left of the two figures corresponds to the road running by the building and shows as a particularly strong (-2.0°C) anomaly on the thermal survey. The feature on the temperature survey is wider than that on the resistivity map, probably due to the former method detecting some of the outer wall of the building in addition to the road to give a combined response. The less pronounced linear feature across the upper part of the two figures corresponds to the wall of the courtyard.

Figure 70 shows a survey over a Celtic temple, also at Verulamium, whose existence has been deduced from resistivity surveys and crop marks. It has not been previously excavated. The data was smoothed with a simple averaging filter. The location of the temple corresponds to that of a cool roughly square feature, appearing as the dark region on the picture.

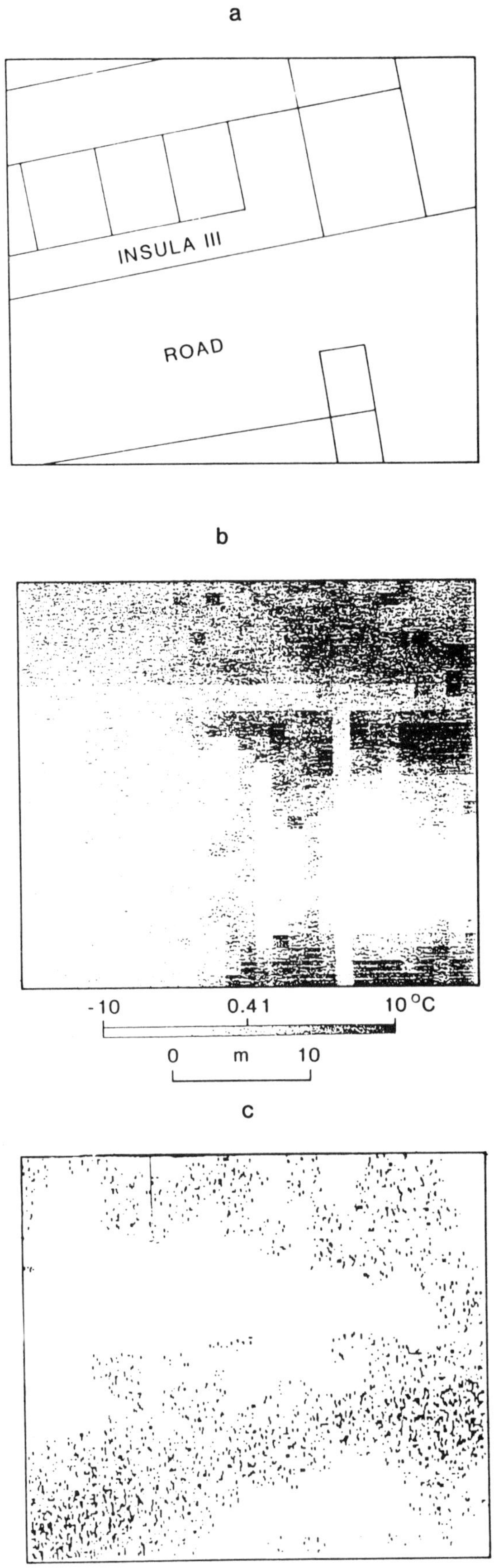

Figure 69 A thermal survey over one corner of Insula III, Building 2 at Verulamium.

a) Location of the survey.

b) The thermal survey data, comprising temperature displacements from the diurnal variation; displayed as a grey scale image of optimal contrast.

c) A resistivity survey over the same area. Res. values range from 35Ω (clear) to 76Ω (black).

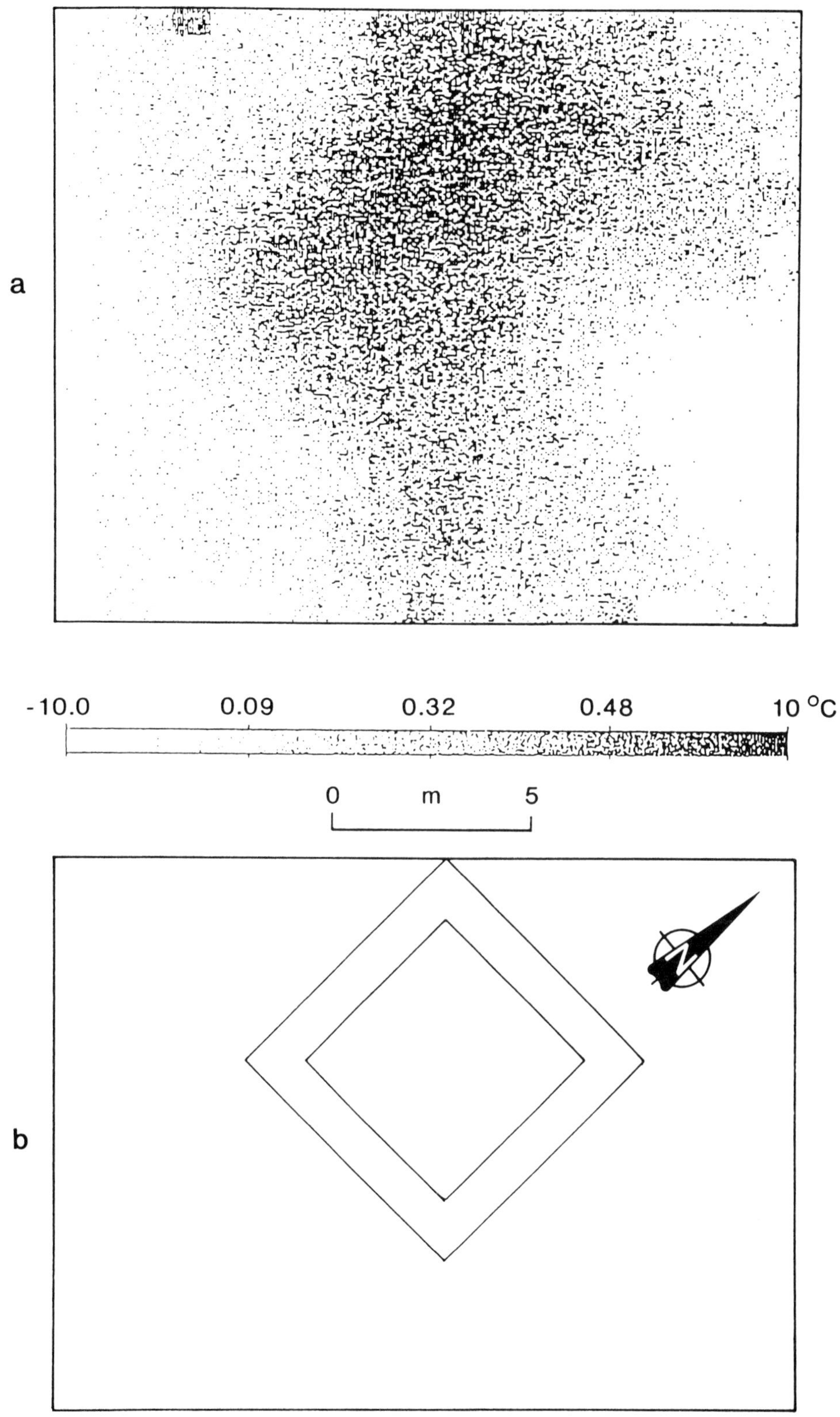

Figure 70 *A thermal survey over a Romano-Celtic temple at Verulamium.*
a) The temperature data smoothed using a simple averaging filter and displayed as a grey scale image of optimal contrast.
b) An interpretation of the above results in the light of information gained from crop marks and resistivity data.

Conclusions

Thermal prospection through the use of direct contact measurements of ground temperature has been shown to be a viable prospection technique which offers some advantages over thermographic methods. The use of subsurface measurements greatly reduces the effect of surface features upon the survey data, temperature anomalies are detectable for much deeper objects in the same medium and the survey equipment is relatively inexpensive, costing less than £1000 for the prototype model.

It would be reasonably simple to construct a data collection system using multiple probes linked to a single data logger or even a set of independent probes with integral, synchronised electronics. With more than one probe in operation at once, a single operator could make measurements at a rate comparable to that in other survey techniques.

The anomalies present in thermal data are relatively easy to interpret. A buried feature will generate a simple region of differing temperature, unlike the complex responses often encountered in other techniques. In addition there is no bias towards any particular orientation of the survey grid as can occur when using resistivity methods. At certain times of the year the signs of anomalies may be simply interpreted in terms of contrasts in thermal properties between the features and the soil.

Further research is needed into the nature of ground temperature anomalies; in particular modelling in three dimensions using expanded computer models and scaled laboratory simulations. With modern microcomputers it should be possible to develop sophisticated image processing methods to extract the information from the survey data and present it in an opitimal form for the archaeologists. It is also possible that the use of meteorological data together with more sophisticated physical models could provide for a more detailed interpretation of the data including feature depths and compositions.

Acknowledgements

We would like to thank M. Pickering for assistance with the computer modelling and C. Saunders, R. Niblett and other staff at the Verulamium Museum for field assistance. Facilities for temperature calibration were provided by M. Povey (Department of Food Sciences, University of Leeds). We are also grateful to W. Manning (Department of Archaeology, Cardiff University) for making available his resistivity data.

References

Benner S.M. and Brodkey R.S., 1984, 'Underground detection using differential heat analysis', *Archaeometry* **26**, (1), 21-36.

Carson J.F., 1963, 'Analysis of Soil and Air Temperature by Fourier Techniques', *Journal of Geophysical Research* **68**, 2217-2232.

Hyndham R.D., Davis E.E. and Wright J.A., 1979, 'The measurement of marine heat flow by a multipenetration probe with digital acoustic telemetry and in situ thermal conductivity', *Marine Geophysical Researches* **4**, 181-205.

Kappelmeyer O., 1957, 'The Use of Near Surface Temperature Measurements for Discovering Anomalies due to Causes at Depths', *Geophysical Prospecting* **5**, 239-258.

Noel M. and Bellerby T.J., 1990, 'A recording soil temperature probe for thermal archaeological prospection', *Archaeometry* **32** (1), 83-90.

Perrisset M.C. and Tabbagh A., 1981, 'Interpretation of Thermal Prospection on Bare Soils', *Archaeometry* **23**, 2, 169-137.

Poley J.P. and Van Steveninck J., 1970, 'Geothermal Prospecting; Delineation of Shallow Salt Domes and Surface Faults by Measuring Temperatues at a Depth of Approximately 2 Metres', *Geophysical Prospecting* **18**, 666-700.

Poley J.P. and Van Steveninck J., 1971, 'Thermal Near-Surface Expression of Shallow Geological Features', *Geophysical Prospecting* **19**, 4, 695-697.

Van Wijk W.R. (Ed), 1963, *Physics of the Plant Environment*, North Holland Pub. Co., Amsterdam.

Part III

The Archaeologist and Geoprospection

The Archaeologist and Geoprospection

Introduction to Part 3

This section derives from a session, held on the final day of the original conference. This part of the conference was conceived as a forum for dialogue between the developers and practitioners of geoprospection techniques and individuals representing tendering field units, researching landscape archaeologists and curators in local authorities. The suggestion was made that not enough information regarding the correct use of geoprospection techniques was available or, indeed, sought. The organisers felt that a structured discussion including viewpoints specific to some of the groups involved would generate more informal expressions of the problem of using geophysics and geochemistry within developer-funded, planning legislation-led archaeology. Whether much ground was made in this direction at the actual conference, was not immediately apparent. However, the greatly increased use of the more 'obvious' techniques as part of the planning process is indeed evident two years on, and perhaps some due can therefore be attached to initiatives and ideas generated in Poole in January 1989.

One concensus that *was* reached during the session was that, if a prospection technique is to be used in such a financially and legislatively constrained environment, then it must be reliable. This includes the reliability of hardware and software, necessary for worthwhile estimates as to the amount of time and money required to do the job, but it also includes the reliability of the actual results themselves, or rather the reliability of their interpretation as provided by the practitioner. The present situation with some techniques of geophysical and geochemical prospection is that they are not sufficiently developed to be used routinely as a part of the planning process. This does not of course mean that they *never* will be, just that, with the present level of research that has gone in to their application, their reliability to produce given results in a given time cannot really be adequately estimated. Other techniques are well understood and have been used for many years, but only have limited application in an environment, such as that of the 'planning constraint' where the need is to locate tangible remains rather than indicate the presence of 'activity areas' and anthropogenic enhancements. The significance and derivation of the latter are often difficult enough for archaeologists to explain to each other, never mind to developers and planning authorities! (although apparently in one Scandinavian country the presence of soil phosphate enhancement attributable to archaeological activity has been reason enough to necessitate excavation to determine whether to divert at least one road scheme (Chris Gaffney pers. comm)). Techniques that this general point applies to include most soil sampling methods, whether for the study of magnetic, phosphatic or elemental enhancements. Within specific research projects, such techniques are often invaluable, especially when the results are used alongside more traditional archaeological data sets (e.g. Bintliff *et al.*, this volume), but there are only very specific instances when they can be adopted as part of the current planning process. Prospection techniques that are now used routinely in the planning process include resistivity survey and magnetometry, and to a lesser extent magnetic susceptibility survey. Both of the former two techniques are fast and their results are reliable, in the sense that, if used, applied and interpreted correctly, tangible archaeology will almost always be identified as such, if it is there. These qualifiers are, of course, of vital importance. No matter how excellent any prospection technique is in theory, it can never be expected to produce good results when used in an inappropriate fashion. Almost all the bad publicity that geoprospection methods have received over the years, the apocryphal stories of the 'boffins' and the expensive 'box' that didn't work, result from bad technique, rather than from bad techniques.

The two papers in this section could not be more different in their starting points, as regards the professional experience of the two authors. Firstly there is a paper written by Michael Aston, a landscape archaeologist with a project that he feels could be greatly enriched through the correct application of geoprospection methods to answer some very specific historical questions. He has a real academic problem, but (self-admittedly) knows too little to really know where to start when it comes to geophysics and geochemistry. This is followed by a paper by Jenny Allsop from the British Geological Survey. She has at her disposal a vast array of remote-sensing and geoprospection methods, many of which she knows could prove excellent in answering archaeological problems, if only she had the data.

Michael Aston's paper discussing aspects of The Shapwick Landscape Project was originally commissioned for the conference itself, but due to delays in the Shapwick fieldwork timetable it was withdrawn from the event, and re-submitted later as an addition to the publication. The paper had been requested specifically because it offered a perspective of landscape archaeology and prospection techniques which might otherwise have not been observed at the conference. That perspective is, of course, one specific to the author, but also in more general terms it represents the 'view' of 'the landscape archaeologist' without the benefit of specialist knowledge concerning the best application of geophysical and geochemical prospection techniques. It was (and is) the organisers belief that this perspective needed to be aired to encourage geophysical and geochemical practitioners to think more in terms of answering specific problems relating to specific archaeological landscapes, applying relevant techniques in a pragmatic manner, sympathetic to local conditions. It is, of course, possible that we all believe that this is what we do anyway, but even if that were true, a good case study remains worthy of informed debate and consideration.

It may be best to suggest at this point that the reader ignores, for the moment, the rest of this introductory piece and turns instead to Michael Aston's paper. Once this is read the paragraphs below can then be viewed in a more informed light. The suggestions made here are, of course, nothing more than editorial comment. They by no means represent *the definitive* geophysical and geochemical approach to the Shapwick project. However, they do illustrate the sort of approach that the editor believes *should* be used in any study of this nature, that being the adoption of a specific suite of scientific applications with the intention of solving pre-determined archaeological problems.

So, what specific aspects of the Shapwick project can be addressed in some way by the application of geophysics and geochemistry?

The basic problem is quite simple. In landscapes such as that at Shapwick, which are predominantly pasture, fieldwalking is not the most useful technique for identifying the presence of archaeological activity. Also, as it is believed by Aston that the early medieval settlement in Shapwick Parish is rather dispersed, and the relevant items of material culture are rather undiagnostic and prone to post-depositional degradation, it is apparent that good fieldwalking results will thus be doubly difficult to obtain. The possibility of using geophysics and/or geochemistry to locate these supposed dispersed settlement sites is an attractive proposition, however, the wholesale application of any such technique would never be realistically possible across a parish with *circa* 800 hectares of potentially rich landscape. Furthermore, Anglo-Saxon agricultural settlement sites are generally rather ephemeral entities, even to the scrutiny of excavation. The possibility that some, or all, of the prospection techniques that could be employed might not detect such features must also be considered.

But a range of techniques are indeed available that might prove fruitful in a situation such as this. The basic requirement is the identification and location of early medieval agricultural settlements. As has been discussed at length in the past, there are very good examples of, if not complete explanations for, the presence of enhanced soil magnetic susceptibility and soil phosphate levels around sites of former human habitation and stock holding (e.g. Tite and Mullins 1971, Proudfoot 1976). A research proposal could therefore be formulated which compared laboratory measurements of soil phosphate and soil magnetic susceptibility levels in samples from locations around known settlement sites, with measurements from transects which bisected areas where fieldwalking and/or documentary evidence had previously suggested the possibility of early medieval activity. As, so far, no actual sites of early medieval farmsteads are known from the parish the test data would have to derive from known sites on similar soils in neighbouring parishes, or from known agricultural settlements of different periods from this parish. It would also be worthwhile to take transects of samples across the one known centre of early medieval activity in Shapwick, that being the old church site. Although the church itself cannot be directly compared to settlements of the type being sought, it is probable that the proposed adjacent 'vicarage' and any further associated structures would have been places where fairly normal domestic and agricultural activities were undertaken, thus providing soil enhancements comparable to those from lay settlements. Transects of multiple samples at, for example 10m intervals across pre-defined target areas and their surroundings compared with similar data from known sites, plus background data from more widely-spaced samples across the landscape as a whole ought to be enough to identify such enhancements *if they existed.*

No mention has so far been given in this example to the two most popular archaeological geophysical prospection methods, resistivity and magnetometer survey. This is precisely because of the nature of the problem being tackled and acts as a good illustration of the tactical application of geoprospection methods to deal with particular situations. The requirement in this example is to locate 'activity areas' where agricultural settlement has previously occurred, as, due to the rather ephemeral nature of the proposed early medieval structures, little in the way of detectable archaeological features is expected. It would be pointless carrying out extensive scanning and/or sampling surveys with a magnetometer, at potentially quite considerable expense, when the level of the expected 'site' magnetic enhancements are not known, and the distinct possibility exists that little other than an increase in background soil magnetic susceptibility across a wide area may in fact be found at any one 'site'. Furthermore low magnetic enhancements over large areas provide their own identification problems. The random use of resistivity survey is even less likely to produce results in this particular example, where few physical structures are expected at any proposed 'site'.

The project design does not exclude these techniques entirely though. Certainly it would be well worth carrying out both types of area survey around the site of the church and its ancillary structures. Here it is already obvious that stone-built structures to some extent survive, and much may be learnt about settlement morphology from any further information. Also any areas where soil phosphate and magnetic susceptibility enhancements were in fact identified could be explored further through the use of magnetometry and/or resistivity survey. Once the probability of a site has been established it is much easier to locate any magnetic or resistance anomalies, however slight, and attach any significance to them that may be dictated by context.

Thus a hierarchy of analytical approaches has been suggested. Starting with the study of soil samples from transects across known agricultural settlements (if possible) and areas of potential early medieval habitation identified from documentary and fieldwalking data, followed by specific studies of sub-surface anomalies within any 'sites' where the enhancements were identified.

This example should serve as both a warning and as encouragement to those wishing to use scientific means to understand archaeological landscapes. Whether or not the suggestions made here are taken up, and whether or not they ultimately prove successful is not important in terms of the wider issues under consideration, although obviously the latter would be of great concern to all parties actually involved at Shapwick! What is important here is the concept that a structured, problem-specific approach to scientific landscape study is indeed possible, and will in general produce far better results than less well structured examples. Furthermore a number of geoprospection techniques are now in such regular use that the creation of projects like the one outlined here need no longer be purely the preserve of the university-based archaeologist. Large-area development proposals require fast and efficient methods of survey where an archaeological constraint is expected or in operation. Geophysical techniques are regularly proving their worth in such an evaluation role. But, if the problems in question could be solved through other methods, such as those outlined here, then, providing a suitably qualified and equipped body can be found to execute the work, there is no reason whatsoever why soil phosphate, magnetic or elemental enhancement data should not be used. Public and private bodies in this country regularly accept scientific data illustrating chemical enhancements as evidence for environmental contamination where public health issues are at stake, surely it would be no more difficult to present similar data as evidence for past cultural activity.

There is no secret to the successful application of geoprospection methods, the only requirements are that the 'tools' are applied in the correct manner by individuals experienced enough to correctly interpret the data.

References

Proudfoot B., 1976, 'The analysis and interpretation of soil phosphorus in archaeological contexts', in D.A. Davidson & M.L. Shackley (Eds.) *Geoarchaeology*, 93-113, Duckworth, London.

Tite M.S. and Mullins C., 1971, 'Enhancement of the magnetic susceptibility of soils on archaeological sites', *Archaeometry* **13**, 209-219.

The British Geological Survey: Geoprospection Techniques Applied to the Archaeological Landscape

J.M. Allsop

British Geological Survey.

Abstract

A variety of remote sensing techniques can be applied to investigate both specific archaeological sites and the paleolandscape. In a three-phased survey programme, these techniques can not only assist in the initial evaluation and assessment of large areas, but also in the detailed survey stage prior to excavation of sites identified during the final analysis. Phase I, normally employed over large "prospective" areas involves the use of vertical black and white aerial photography in combination with airborne multispectral scanner data with image enhancement. These data can be used to identify both specific archaeological features undetected by conventional aerial photography and areas most likely to contain preserved archaeology.

Phase II, concentrated on areas identified by the airborne surveys, requires more detailed investigations by rapid ground remote sensing techniques such as EM conductivity. Enhanced computer software can provide 3-D colour displays of the subsurface to 7m. The detailed surveys involved in Phase III can equally be applied to smaller urban sites, geology permitting. A multicomponent survey can be devised from a variety of techniques to provide the most suitable combination for the detailed analysis of the site prior to the planning of the final excavation stage. In a similar way, a number of techniques are available for marine and near shore surveys to locate and map drowned settlements and paleolandscapes.

Back up facilities including lab. soil tests and data held in the National Geosciences Data Centre are an integral part of the planning, execution and interpretation of these data. This methodology can provide a valuable tool for the archaeologist involved in assessing 'threats' from development and mineral extraction which are liable for 'developer' funding.

Introduction

The British Geological Survey (BGS), whose principal role is the systematic surveying of the UK landmass and its adjacent Continental Shelf, employs a number of field- and laboratory-based techniques that can be applied directly to archaeological investigations. The BGS is the largest component institute of the Natural Environment Research Council (NERC) which also has facilities of use to archaeologists. Many of these techniques are becoming accessible to archaeology, especially with the expansion of rescue archaeology and, in particular, where developer funding is involved. This type of funding, which formed about 17% of the total archaeology funding budget in 1988 (Plouviez 1988), can be negotiated at the planning stage to include prospection surveys whether the site is urban or rural. Detailed prospection surveys, using three or four methods based on different physical or chemical parameters, can provide useful data about the site and a guide for the archaeologist in decision-making about trial trenching. Developers benefit by ensuring start dates for contractual work, which may even be brought forward in some cases. By such means, understanding and trust between developers and archaeologists are fostered.

A recent conference jointly organised by SCAUM and the British Building Federation on 'Archaeology and Redevelopment' (12 May 1989) included suggestions that developers should contract a geophysical survey prior to planning so as to assess the site (perhaps as part of an engineering geology site investigation); at a later stage the results might be made available to archaeologists in their planning of rescue work. The success of the Code of Practice, as outlined by the Developers and Archaeologists Liaison Committee (affiliated to the Institute of Field Archaeologists, IFA) was also discussed. The ideas put forward by the conference regarding geophysical surveys in urban rescue situations would certainly go a long way towards realising the suggestion (Mellars 1988) by the Council for British Archaeology (CBA) that "the IFA should take a positive role to ensure that in a situation where developer funding under contract becomes increasingly frequent, archaeologists should uphold professional standards of information retrieval and dissemination and ensure concomitant resources" (Cleere and Marchant 1988).

The following paper outlines the main prospection techniques, together with the extensive 'back-up' resources, available within BGS and NERC; they would help in the type of situation where essential, adequate funding has been provided by the developer.

Remote Sensing Techniques

Although in most geological contexts remote sensing in the strict sense only refers to landsat and airborne scanner analysis, this section of the account also includes geophysical surveying and engineering geological investigative techniques.

The Hart report (SERC 1985) suggested that the "development of ground radar and shallow seismic reflection techniques may be profitable" and "that satellite imagery may also prove invaluable". However, these techniques have been slow to develop in archaeological work; in fact, in many situations new techniques have evolved which are equally important.

Landsat and Airborne Scanner Imagery

Satellite and airborne remote sensing data and the relevant interpretative expertise exist within NERC Scientific Services (NSS) and BGS. This is a new concept in science-based archaeology and a valuable tool in the assessment of larger areas for settlement and landscape studies. This type of remote sensing, together with stereo-aerial photographic interpretation has been used for large scale geological mapping and mineral reconnaissance surveys abroad. The group responsible, based in the Overseas Directorate of BGS, holds regular courses in photogeological interpretation as part of the UK Aid Programme provided by the Overseas Development Administration.

The bulk of the Landsat data (with a 30m ground resolution) is held by NSS on a digital tape system - gaps in the coverage exist only because of adverse weather conditions in northern Scotland. The cost of acquiring such data from NERC would be at the cost of reproduction. However, a copyright system exists which may require either NERC or BGS personnel to request and interpret these data on behalf of the archaeologist. The application of Landsat and airborne scanner data to archaeological problems would be a logical extension of the work of the Remote Sensing Development Applications Unit of NSS. It may, in this case, need some input from the NERC Unit for Thematic Information Systems based at the University of Reading.

In the case of high resolution remote sensing data, often required for smaller archaeological sites, the NERC Airborne Scanner System may be more applicable. Since 1982, NERC aircraft have used the scanner system on numerous sites, following up specific requests by geologists and other scientists. This system has the advantage of recording data from the visible to the near infrared and thermal infrared ends of the electromagnetic spectrum, the thermal infrared may well be invaluable in the case of, for example, medieval burial grounds. Although very similar to Landsat data, the scanner system works on 11 spectral channels (7 for Landsat) with a higher ground resolution down to 2m. Many areas have now been recorded, usually in 10 by 10km cells. Some areas of interest to archaeologists could well have been previously scanned.

Colour Plate 1 (Overleaf, upper) *Image obtained from the NERC aircraft over the Fenlands revealing ancient, natural drainage systems. Here, features characteristically composed of clay or loam contrast with the surrounding peat and clay (After Lynn, 1986).*

Colour Plate 2 (Overleaf, lower) *Colour graphics computer display of ground magnetic data from a site at Melton Mowbray, N.E. Leicestershire. The remaining portion of a Roman coffin (located during a watching brief) at 32E and 32-39N in a large ditch (blue- low magnetic values). The highest magnetic values (red) indicate the stone platforms of nearby Medieval houses belonging to the deserted village of Sysonby (excavated 1988). An amateur dig is planned to recover the remaining parts of the coffin on the basis of preliminary geophysical surveys. (After Allsop 1989 a).*

Again it may be cost effective for NERC-personnel, with the expertise, to order and interpret these data for the customer. The NERC aircraft is also capable of taking straightforward aerial photographs, but this seems to be "in hand" in most County Archaeological Units.

A new piece of NERC equipment available on loan, the spectro-radiometer, could enhance the detailed interpretation of this type of remote sensing imagery. This provides additional spectral information from visible to infrared regions of the electromagnetic spectrum to assist in the detailed interpretation of remotely-sensed data from aircraft and satellite recordings.

A recent development in the use of remotely sensed data has been research into the relationship between vegetation and geology; this technique would be a useful tool in many archaeological situations. Using image analysis enhancement, individual fields can be studied from NERC scanner data (mainly thermal, middle and near infrared information) to detect subtle changes in the surface soil characteristics. Features such as disturbed ground (grubbed out woodland, hedges etc. and filled in ditch systems) can be located together with "natural" soil changes, for example, clay, sand and loam, together with changes in soil moisture and small scale topography.

On a larger scale (Colour Plate 1) NERC scanner data, combined with image analysis facilities, can reveal ancient drainage systems in the Fenlands - this may be invaluable in constructing the past landscapes and drainage patterns which governed earlier settlement patterns (Lynn 1986). Such image enhancement of scanner data may be equally important to landscape studies in similar areas, such as the Somerset Levels.

Ground Geophysical Surveys

Most geophysical techniques are applicable to archaeological problems. However, according to the type of archaeological site (rural or urban), the final degree of preservation required for the site and the subsurface geology and soil conditions, each multi-component survey has to be tailored and adapted to local requirements. The usual strategy would be to use three, or even four, methods - each based on different physical parameters- to ensure all types of features on a site have been adequately located. All surveys involve, where possible, digital data capture in the field and the appropriate computer software to enable the rapid production of final results (often in 3-D) - a factor essential to most commercial requirements and also to archaeology in situations immediately prior to rescue excavations. The trial trenches could be planned to encompass all types of geophysical signatures recorded on the site; establishment of the cause of each type of signature could remove the need for total excavation of a site, depending upon its significance or the extent of the excavation in the time available.

44
43
42
41
40
39
38
37
36
35
34
33
32
31
30
29
28
27
26
25
24
23
22
16
18
20
22
24
26
28
30
32
34
36
38
40

General geophysical surveys can include a combination of many different methods, each utilising the most suitable equipment for the site concerned. These include resistivity (with various array options), Induced Polarisation (IP), magnetometer and gradiometer, electromagnetic (EM), VLF(Very Low Frequency) -EM and VLF- resistivity surveys (the latter method has proved the most successful technique for locating subterranean caverns), ground radar and shallow seismic surveys. In addition, laboratory analysis of physical properties of materials and soils can enhance the interpretation of a survey, assisted by a comprehensive data-bank held in BGS.

The EM technique is also a rapid method of geophysical prospection in which penetration can be varied according to the situation. The EM 31 equipment (an average of 500 readings per day) can be adjusted to different depths of penetration by changing the distance of the boom from the ground surface and by changing the plane of orientation of the coils from vertical (shallow) to horizontal (deep). Used in conjunction with the EM 34 equipment (variable length of boom related to depth of penetration), a conductivity "pseudo-profile" can be constructed providing a section through multi-layered drift deposits (Cornwell 1985). When this is calibrated with augered soil samples, tested in the laboratory, an interpretation of the variation in layers of gravel, clay, alluvium etc. can be made. This has been successful in providing subsurface data for the BGS mapping programme (Zalasiewicz *et al.* 1985) in areas of drift deposits in East Anglia (Mathers and Zalasiewicz 1984; Zalasiewicz & Mathers 1986). The use of the EM technique in prospecting areas of sand and gravel deposits could be directly applied to archaeological prospection, as in the Castle Donington area (Clay 1986) where more archaeological evidence might have been retrieved from the advancing gravel extraction and the team could have concentrated on the most promising areas of archaeological interest. This type of specialised survey (Mathers and Zalasiewicz 1985; Zalasiewicz and Mathers 1985) could play an essential role in areas under threat from mineral extraction where the Code of Conduct is out of date and in need of urgent revision (Chadwick 1988).

Many geoprospection results are still processed and displayed as dot-density or contour maps (originally noted by Aspinall et al. 1983), for example, overseas archaeological reports presented at the United Kingdom Geophysical Assembly and other similar geophysical conferences. BGS has developed advanced computer software for the recording and processing of digital field data. The final data plots can be displayed in a variety of ways using modern computer colour graphics (Colour Plate 2). In many cases image analysis of data sets so gathered can be employed to extract more detail from surveys and perhaps, provide additional information for the archaeologist. The image analysis system allows a comprehensive analysis of different data-sets, for example, comparison of geophysical results with geological and geochemical information (where available). Software including multivariate analysis can later analyse the correspondence between different data-sets such as geophysical surveys and excavation results.

Field Example

In the Spring of 1989 the author was invited to carry out geophysical surveys on the site of a proposed Iron Age farmstead project at Donington le Heath, Leicestershire (SK 420 126), adjacent to the Medieval Manor House (Leics. Mus.,Arts & Records Service 1989). The purpose of the survey was two-fold. Firstly to ensure that later excavations, during the construction of the outer ditch and house foundations (based on a plan of an Iron Age site previously excavated at Enderby), would not destroy other significant archaeological remains. Secondly, the use of various types of geophysical surveys along the line of the boundary ditch could be directly compared with the excavation results to a depth of 1.5m.

The selection of geophysical techniques was made to illustrate the use of several methods based on the measurement of different physical parameters. In this case, a detailed comparison of these techniques with the soil profile should show features detected by one, several or all the methods employed.

The initial geophysical surveys were completed using a conductivity meter (EM 31 - deep and shallow soundings), gradiometer and magnetometer. Later, these profiles were compared with the excavation results (including soil removed during the dig and with the remaining soil profile). Figure 71 shows the proposed trench-line in relation to the Manor House during the geophysical survey (a) and the same view along the trench during the post-excavation comparative studies (b). Using the EM31 technique, an alternating electric current passed through one of the two coils, produces an alternating magnetic field. The soil responds by becoming partially magnetised and induces an electrical current in the second coil. The alternation of the initial (injected) field allows the instrument to measure the response of the soil and the way in which the induced magnetic field decays once the injected field is removed. The amount of magnetisation induced in the soil gives the magnetic susceptibility; the time taken for the field to decay indicates the electrical conductivity of the soil. Therefore two surveys are being completed at the same time. In addition a deep and a shallow reading is taken (by changing the plane of the coil from horizontal to vertical, respectively) which can build up a depth profile of changes in soils and features at different levels below the surface.

The results from the fluxgate gradiometer survey are shown as a series of positive and negative anomalies, both disposed about an arbitrarily chosen zero. If the zero represents a region of "no magnetic feature", then small positive anomalies may be identified as pits, ditches etc., containing humic materials in a less magnetic matrix. Small negative signals may be caused by non-igneous stone features buried in "natural" soil. However, by the nature of magnetic anomalies, "positive" features may have associated negative signals in their vicinity; the converse is true for "negative" features. The presence of large positive signals with a spread of, say, 2 metres may well be identified with fired features such as kilns or hearths; highly localised large positive and negative signals can be attributed to buried iron artifacts, but in these cases readings can be erratic.

Figure 71 *Geophysical surveys at Donington le Heath, Leicestershire.*
a) *EM conductivity survey prior to excavation of boundary trench. (Medieval Manor House in background).*
b) *Later examination of the trench after excavation to 1.5m, correlating man-made and natural features with anomalies identified by the geophysical surveys.*

A total magnetic field profile was also measured. In this case, physical disturbance of the ground often produces a corresponding and measurable disturbance in the earth's field; for example, ditches penetrating the sub-soil may later silt up with a mixture of topsoil, which, in effect, locally deepens the top soil. The magnetic content of topsoil, which is often far greater than that of the sub-soil, can be further enhanced by activities of human occupation such as burning. Clays, baked in a kiln, acquire a strong remnant magnetism on cooling and provide a strong and usually clear detectable signal.

Most archaeological features produce only a small local enhancement in the earth's magnetic field; anomalies in the range 5 to 30 nanoteslas (nT) are typical; these have to be detected against the background of the earth's field of about 50,000 nT at British latitudes. The detectable fields of kilns, furnaces etc. are often far greater than that, typically 1000 to 5000 nT.

The influence of geology on a survey is a major consideration, especially if the parent rock is igneous or the drift deposits contain an igneous element. Donington le Heath is in an area of glacial sand and gravel overlying the mudstones of the Mercia Mudstone Group (Triassic). The glacial deposits are variable, the proportion of gravel to sand may vary from 5 to 50 per cent and up to 25 per cent of the deposits may be silt. The gravel fraction usually consists of hard "Bunter" quartzite, with small quantities of porous sandstone and occasional boulders of local igneous rocks from nearby sources of diorite (Worssam and Old 1988).

Over the remaining survey area the EM31 equipment was used to "scan" by walking and observing the meter to check for the significant anomalies. A short magnetometer and EM31 traverse across the proposed location of the large round house produced no sign of anomalies likely to affect the later work of house construction.

The results of the four geophysical surveys are shown in Figure 72; each feature has been labelled and corresponds to the final interpretation in Table 4. The description of the likely source for each geophysical anomaly, based on excavation results and field examination of the finished trench, also shows the differing responses of the geophysical techniques. The EM31 surveys indicate areas of high conductivity associated with "stoney", well drained areas; the gradiometer and magnetometer results indicate areas (positive anomalies) where there is a higher proportion of silt, a depletion in the stone and gravel content of the soil or the presence of igneous rocks. During the field examination, magnetic susceptibility measurements were made on the large boulders (removed during excavation from the area of the "wall" near the Manor House and the area between stations 15 and 19). The boulders, up to 0.4 metres in diameter, consisted mainly of local igneous rocks (diorite and granodiorite) with a magnetic susceptibility of 8.9 to 9.7×10^{-3} cgs units; a few samples of well-bedded, micaceous sandstone boulders produced values of 1.9 to 2.4×10^{-3} cgs units. Although these boulders may have originated from the local glacial deposits, they appear to have been grouped into two areas of the trench and are, therefore, not a naturally occurring feature of the local deposits.

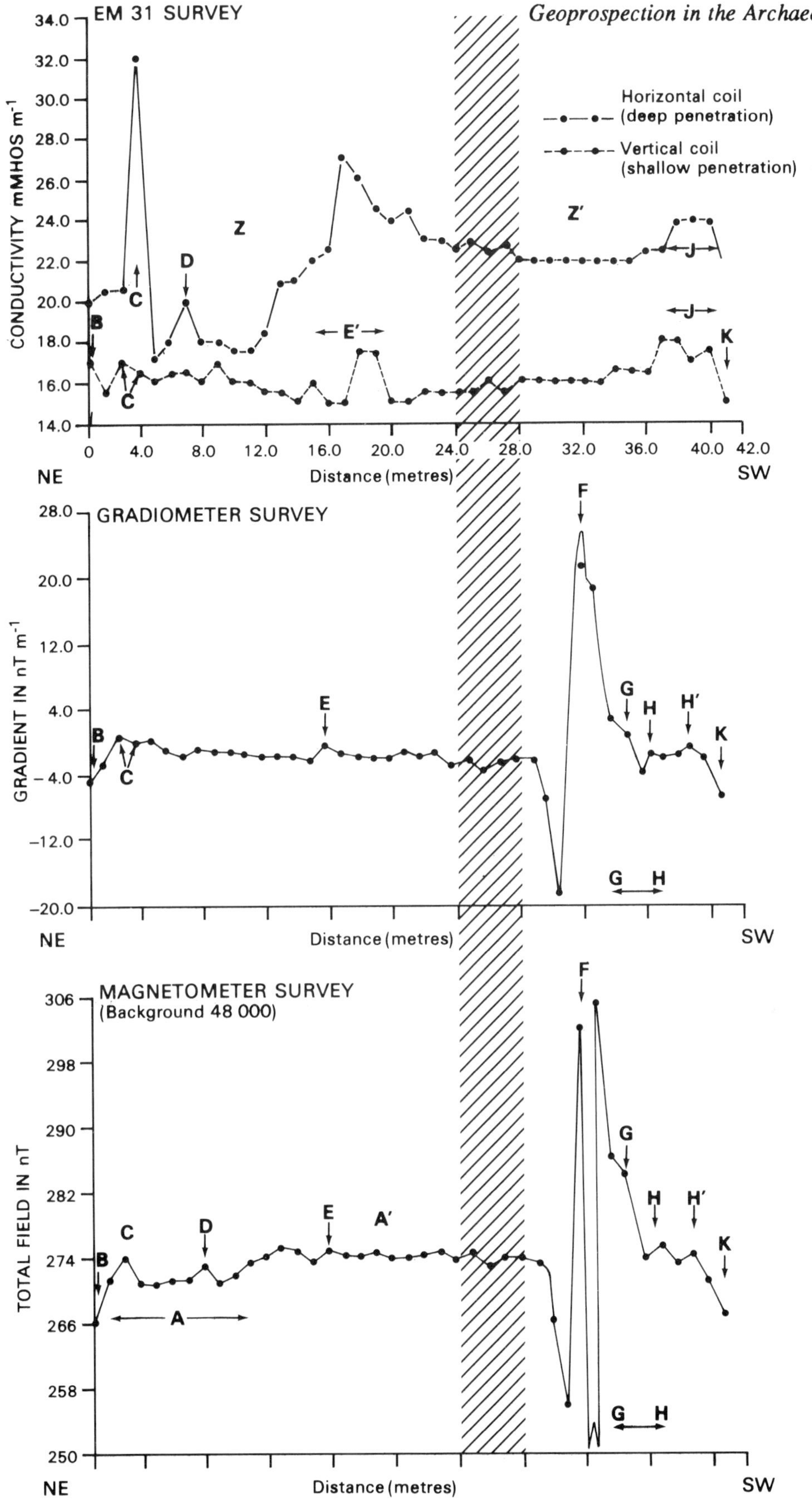

Figure 72 *Geophysical survey results along line of trench at Donington le Heath compared with the results of later excavation. The anomalies from EM 31 (deep and shallow), gradiometer and magnetometer surveys react in different ways to soil changes and archaeological features, illustrating the need for multicomponent surveys (See Table 4 for key and description of labelled features).*

The results of the traverse clearly show the need for using at least three or four different geophysical techniques to ensure that all the significant archaeological features can be detected (Table 4).

Marine Geophysical Surveys

Marine geophysical surveys, either near-shore or in deep water, can be carried out using magnetic techniques together with sidescan sonar, pinger and boomer equipment. These could be used in the mapping and detection of in-shore drowned settlements and submerged landscapes. In shallow water, BGS geological divers collect sediment samples for laboratory testing in order to make more accurate interpretations and assessments. Joint academic studies have, in the past, modelled the glacio-marine depositional environment around British shores.

The ocean-going Research Vessels provided through NERC Scientific Services are fully equipped with up-to-date computer software, modern navigational and positioning systems. The problems often encountered by marine archaeological surveys (Fleming & Redknap 1984; Joyce 1985) could well be assisted by the standard equipment of these research vessels in combination with sonar or marine geophysical surveys (for example, magnetic) in locating such features as wrecks. NSS (NERC Scientific Services) Vessels are available for charter at certain times of the year depending on their scientific cruise programme.

Various NERC equipment pools, including geophysical equipment, which provide equipment for specific projects within University departments and NERC Institutes, could also be invaluable to larger scale archaeological surveys. A full inventory of this pool, together with the (land) geophysical, geochemical and other facilities within NERC and its component bodies (including BGS) is to be published as an Appendix to the report of the sub-committee on the Funding of Applied Science to the Forum for Co-ordination in the Funding of Archaeology. However, as mentioned in the report, certain items listed in this Appendix may not be generally available to archaeologists or, as in such cases as NERC Research Vessels, availability is subject to the NERC timetable.

Minor methods of Prospection

Finally, photogrammetric studies can be used to study the stability of monuments such as Dunbar Castle (Gostelow 1988); a technique usually applied to landslip and slope investigations, but equally applicable in an archaeological context. Buildings of interest under threat from minor shocks, especially in areas of previous mining activity, can be studied using seismic monitoring - this is another adaptation of a normal geophysical technique used in site investigations, usually where dams or similar constructions are threatened.

Feature (Station No.)	EM31 (deep) Mmhos/m	EM31 (shallow) Mmhos/m	Gradi-meter nT/m	Magnet-ometer nT	Interpretation
A (0-11)	-	-	-	low (c.270)	Acid, sandy soil very wet (orange colour)
A'(11-end)	-	-	-	high(c.275)	Dark, loamy soil (dryer)
Z (0-14)	Ave.19	-	-	-	Soil becoming generally more stoney
Z'(14-end)	Ave.23	-	-	-	More near-surface silt
B (0-1.5)	-	+ve	-ve	-ve	Start of ditch,(?moat)
C (3-4)	+ve(12)	+ve(2)	+ve(3)	+ve(2)	V-shaped ditch- silt (70cm deep), but very stoney below
D (6-7.5)	+ve(2)	-	-	+ve(2)	Flat-bottomed gully 1.5m wide, silty but more stoney at base of trench
E (15)	-	-	+ve(2)	+ve(1)	Area of larger stones (?igneous)
E'(15-19)	+ve	+ve	-	-	Area of larger stones (silt content same, fewer smaller stones)
[**24.1-27.9**	WALKWAY - NOT EXCAVATED]				
F (c32)	-	-	(-20 to+24)	(257-348)	Silty V-cut ditch
G (c.35)	-	-	+ve	+ve	Ring of large igneous boulders (matrix remains "normal")
G-H	-	-	-ve	-ve	Stony gulley (thin silt layer above)
H (36.4)	-	-	+ve	+ve	)Two areas of near-)surface burnt
H'(38)	-	-	+ve	+ve	)material (c.4cm)
J (c.37-40)	+ve	+ve	+ve	+ve	"Wall" - numerous large stones (igneous) coincident with gully
K (40.5)	-ve	-ve	-ve	-ve	Start of deep ditch

Table 4 *Donington le Heath - Identification of geophysical anomalies with proven features (see Figure 72)*

The Global Seismology Unit, based in Edinburgh, are building up a data bank of historical archives and documentary references for earthquake occurrences and volcanic activity. This can lead to a reassessment of the magnitude and epicentre of events. These events may be responsible for local reports in historical documents, such as parish registers, as in the case of the Colchester earthquake of 1884 which revealed that the epicentre was nearer the village of Peldon, south of Colchester (after-shocks have also been recorded in documents) and that the magnitude of the main shock was about 4 to 5 on the Richter scale (Allsop & Smith 1988). Records are also being compiled, both in

Britain and in Europe, to assess more widespread events. These archives may be of use in the future, should important historic buildings be under threat. Alternatively the seismic history of a locality may be vital evidence if a survey is required to establish the stability of an archaeological site.

Proximal Prospection Methods

These methods may require sampling the ground or even penetration by narrow-gauge drilling (borehole) or augering techniques.

Engineering Geology Site Investigation

Engineering geological site investigations are often undertaken prior to contractual work or at the planning stage for redevelopment sites or, on a larger scale, dam and motorway construction. The assessments, often in conjunction with geophysical surveys, are vital to the stability and, in some cases, the safety of a construction. The engineering geologist also has the advantage that any of the findings, during the prospection and assessment stage, are later proven by excavation of foundations during the site development and, therefore, the link between the survey and the reality of the site provide a special kind of experience and knowledge. New techniques including cross-borehole geophysical surveys have increased the detail and amount of subsurface information, on which vital engineering decisions are often based.

Landslips are common and may involve historic monuments. The potential for slippage, the likely conditions for its occurrence (i.e. water content of the soil and the geotechnical properties of the soil) and any history of previous slippage are important. In the case of Hadleigh Castle (Hutchinson and Gostelow 1976) this type of survey was carried out and the authors concluded that a major landslip in the cliff had partially damaged the castle soon after its construction in the 13th century.

Soil Prospection Surveys

Soil prospection surveys can be carried out in three ways - "on-site" field analysis as an initial "rough" guide, detailed laboratory tests on field samples (either from traverses or from spot locations) and, finally, detailed tests on narrow gauge auger samples taken from a suspected settlement site or from the engineering geology site investigation borings on urban sites. Laboratory analytical techniques, as applied to field or core samples are similar; however, on-site techniques are more limited and most field methods do not provide absolute data values.

In the field, tests for "total" phosphate content can identify areas of high phosphate content, although factors such as geology, soil conditions modern (and older) farming practices have to be considered before the identification of a possible settlement site can be considered likely. Changes in subsurface geology can be equally important - limestones, clay content and the acidity of the soil (in particular sandy soils) can alter the distribution by leaching and transport of phosphates out of the soil profile. The water content of the soil at the time of the survey is also an important factor. Clearly, it is important to establish a detailed knowledge of the geology, soil geochemistry and hydrogeological properties of the soil prior to any archaeological investigation.

Laboratory tests on soil samples from traverses or cores provide extremely useful information. Tests for carbonate and sulphate content, percentage of organic material, acidity, magnetic susceptibility and more detailed phosphate analysis, together with particle size and sedimentation analysis provide a better basis for assumptions and interpretations about possible settlement sites. The methods of phosphate analysis, in this case, would provide values suitable for detailed interpretation rather than the high, medium and low estimates provided by on-site techniques. Total phosphate values may be misleading, since the "available" phosphate is mainly the concern of the archaeologist - that is, the phosphates present in the soil which can be absorbed by living plant material. Trace element analysis (including metals) may also be an important test; these can give an indication of metal working on a site, provided weathering of the subsurface rock has not produced these elements naturally. However, a knowledge of local background levels and the dynamic nature of the natural processes which have occurred since the occupation of the site, such as changes in water level or surface water courses or the presence of sub-terranean water sources, can also be vital to any final interpretation. Hydrogeological tests using tracer-tracking techniques are a standard technique and have already proved important to archaeologists. Studies such as those completed on the Mucking "silhouette", are invaluable to hydrogeologists, since transport models based on a known time-scale from an archaeological site can assist geological research into hazardous waste disposal (Thomas *et al.* 1988).

In the case of small-bore core samples, which, for the most part, would be derived from urban engineering geology site investigations, there are additional types of analytical techniques available. Buried soils can be analysed for clay content - clay mineralogy is an important factor in the consideration of water movement. The soils can be examined for pollen, macrofossils (arthropods being particularly significant) and microfauna. The palaeoenvironment and depositional history of the soil profile can be interpreted. This is similar to the powerful chronostratigraphical tool available for post-excavation work. Such studies can define interglacial sediments of the Palaeolithic (identification of artifacts), the clearance of land in the Mesolithic period and farming practices in the Neolithic. Settlement sites can also be identified using these additional techniques in conjunction with standard soil sample analysis.

Preliminary background research into the geology, soil survey records and the geochemical data bank would be essential before any detailed studies could procede. An integrated multidisciplinary approach to soil prospecting could then be provided as a "package" involving geologists, geochemists, geophysicists, mineralogists, biostratigraphers and other specialists. Such a system could also be adapted according to the "customer's" needs for post-excavation work.

Geological Surveying, Back-up facilities and Environmental Studies

Regional geological mapping, in association with sediment analysis and petrology, can assist archaeologists in recreating past landscapes especially in non-glaciated areas such as the South Downs of England (Bell 1983: Allen 1988).

Detailed geological mapping, combined with geophysical surveys and various back-up facilities within BGS may also be a valuable prospecting tool for landscape studies and the definition of patterns of settlement. Modern mapping techniques, assisted by previous notes often dating back to the last century, can help in defining environments of deposition and palaeoclimates. In Lincolnshire, detailed mapping has defined the difference between Roman and Medieval saltern mounds and, therefore, suggested the likely trade routes which may have existed at these different periods (Pattison and Williamson 1986). Landform interpretation, assisted by techniques described in the soil prospecting section (to establish ancient river courses and terraces), were used to reconstruct the palaeoenvironment at the time of the burial of the Brigg Raft (Fletcher 1981).

Finally a geophysical experiment was carried out in the Redditch area, using augering control in combination with resistivity, EM, Induced Polarisation, seismic refraction and ground radar surveys, to assess the success of each technique over scattered drift deposits (Clarke *et al.* 1982). The exercise was intended to determine the best combination of techniques, in addition to a preliminary airborne scanner survey (Hamblin and Crofts 1984), to assist field mapping geologists in the future.

Combined local geological knowledge with other specialist techniques within BGS can be essential before prospection surveys, excavation planning or simply fieldwalking.

In areas of previous mining activity, such as N.E. Leicestershire, areas of surface ironstone extraction may often be reclaimed and put under the plough. After many years, and without detailed knowledge from BGS records, fieldwalkers may recover finds belonging to the Roman period or older. However, near the Roman town of Goadby Marwood in N.E. Leicestershire, the reconstituted land was infilled indiscriminately. The "finds" do not relate either stratigraphically or spatially to the Roman settlement and, without prior research into archive material, the finds may be misleading and a waste of valuable time and effort.

Archaeological petrography combined with local geological knowledge may also be important. The work by Collins (1979) assisted archaeologists to develop the idea that the early inhabitants of Skara Brae had a self supporting economy.

A similar problem has been encountered in N.E.Leicestershire. Part of a Roman coffin, made from sandstone, was recovered during a watching brief near Melton Mowbray. The stone had been worked in a similar fashion to the limestone equivalents found elsewhere in Leicestershire. No similar sandstone working from this period is known in this area; local limestone sources are convenient and widespread. Petrographical work identified a source from the north; it also suggested that the outside of the coffin had been pounded flat before enigmatic incisions had been added. Physical properties measurements provided a strong magnetic susceptibility reading compared to the general chalky boulder clay assemblage on the site and so the magnetometer surveys revealed the location of the coffin for later excavation (Colour Plate 1).

A simple desk-study of geological maps can often provide a sound background for the identification of settlement sites prior to prospection. The significance of small patches of surface gravel (or in the near-subsurface) in an area of boulder clay can indicate the presence of previous settlement sites. These patches may often have controlled the later nucleation of settlements and the modern extent of village expansion. In 1872 the Public Health Act made strict regulations regarding villages built completely on boulder clay, forcing the sealing of all unpotable water sources and the building of reservoirs to supply drinking water. This caused a great debate in many scientific journals on the relationships between geology, water supply and village locations (Hull, 1876). Water supply is essential; combined with well-drained ground for building it can be an overriding factor in early settlement patterns. On Lias soils, settlements may relate directly to thin limestone or sandstone beds within the water-logged clays, especially where these aquifers relate to higher ground. Villages will be confined, even today, to the extent of these bands as a source of water (from wells). Hence the shape and size of villages are often controlled by topography and the length or width of the aquifer.

In addition to previous field mapping notes, manuscript and detailed versions of geological maps held in the National Geosciences Data Centre (NGDC), well records can also provide clues to known settlement sites which do not conform to the pattern outlined above. Well records may often show that the aquifer is just below the ground surface - either as a gravel "lens" within the boulder clay or a limestone bed in the near-subsurface which may appear on a surface geological map as Lias clay and therefore mislead the landscape archaeologist.

The geological mapping of the UK and offshore Continental Shelf is ongoing, but started in the early part of the last century. The surface geology is an important factor which must be investigated before any type of reconnaissance survey is initiated.

Within the NGDC the manuscript records and field notes of past geological mapping may provide details which are not necessarily included on modern published maps, often for reasons of scale. These records, dating back to the 19th century, can provide extremely useful or even vital evidence. The centre of Melton Mowbray, Leicestershire, appears to be located, from published geological maps, on an area of chalky boulder clay extending down to the recent river alluvium in the valley of the River Eye. However, the notes of the 19th century geologist suggested small patches of gravel which did not appear to be continuous and, therefore, were not included on the master copy. The discrepancy between the early settlement history of the town, its flourishing market at the time of the Domesday survey and the apparently unsuitable ground conditions for settlement was solved by recent recordings of soils during redevelopment of the inner town and two excavations within the possible Saxon "core", near the church. It appears that two gravel river terraces exist above the flood plain, explaining the ability of early settlers to build with safety on well drained land near to a river crossing which has been significant at least since Roman times.

Therefore, the NGDC and the BGS Library facilities can provide a significant source of information to the archaeologist.

NERC Scientific Services also provide important facilities in support of BGS activities. In addition to VAX 8600 software developed specifically for processing and interpreting digitally recorded geophysical field data (probably requiring little adaptation to the relatively small-scale studies associated with archaeological surveys), NERC Computer Services have a variety of highly developed software at specialist workstations. In particular, the Intergraph work station and the image analysis systems available at many NERC sites around the country are potentially very useful.

Computer databases are also an integral part of BGS facilities, including physical properties and a large geochemical data-bank covering the whole of Great Britain. The latter can be used in conjunction with image analysis facilities and to form the basic ground work for geochemical investigations of trace metals and soil sample studies, comparing these with local background levels.

The Need For An Interface

Archaeologists may require an interface to discuss general problems before more detailed plans are considered. Since the BGS initiative into science-based archaeology, "user-friendly" contact points have been established in the Aberystwyth and Edinburgh offices, in addition to BGS Keyworth (Allsop 1989b). The "customer" archaeologist may also benefit from working alongside the specialist during the final interpretation. At a specialist computer work station, the results of the survey may be enhanced (perhaps presented in 3-D with colour graphics) so that more information (directly related to the original archaeological problem) may be gained than would have been possible using conventional display methods.

Conclusions

Over the past few years the applicability of a whole new range of scientific techniques in archaeology has been recognised as of increasing importance. This is especially true because of the emergence of rescue archaeology and, in particular, developer funding. So as to maximise useful data it is essential to inform archaeologists and developers alike of the types of useful research tools available through NERC and BGS. The additional survey costs can then be discussed within the initial funding negotiations.

In most archaeological investigations a combination of the prospection techniques and back-up services can be adapted to each individual situation to provide a comprehensive 'best appraisal' of the total features on a site, many of which could remain undetected by a less comprehensive survey. As a result, developers and archaeologists may benefit from this type of package, despite the initial cost. The archaeologists receive a detailed record, especially in cases of partially excavated sites, and the developers ensure the time limits are adhered to or even shortened.

While NERC does not have a direct funding role for science-based archaeology, it does have a range of resources which have either not, as yet, been applied to archaeology or do not exist elsewhere. Therefore the potential exists for NERC to supply such services and, at the same time, avoid the possibility of parallel research duplicating already well-established techniques (Pollard 1988).

Acknowledgements

I am grateful to David Lynn (NERC) for providing the airborne scanner data and Mike Raines for his assistance during the field work. Patrick Clay and Ian Hind (Leicestershire Archaeological Unit) also provided valuable assistance during the Donington-le-Heath field survey. Finally, I would like to thank the Director (BGS), Chief Geophysicist and Heads of the Regional Geophysics and Deep Geology Research Groups for their support and advice. This paper is published by permission of the Director of the British Geological Survey (NERC).

References

Allen M.J., 1988, 'The archaeological and environmental aspects of colluviation in south-east England', In W. Groenman-van Waateringe and M. Robinson (eds.), *Man -made soils*, BAR International Series **410**, 67-92.

Allsop J.M. and Smith N.J.P., 1988, 'The deep geology of Essex', *Proceedings of the Geologist's Association* **99**(4), 249-260.

Allsop J.M., 1989a, 'What can NERC offer the archaeological community?', In L. Jones (Ed.), *NERC News* **9**, 15-18.

Allsop J.M., 1989b, 'The British Geological Survey Initiative - A Service to Archaeology', In M. Pollard (Ed.) *Science-based Archaeology Newsletter* **2**, 2-3.

Aspinall A., Dockrill S.J. and Pocock J.A. (Eds.), 1983, *Geophysical Surveys 1982*, Occasional Papers No.3, Department of Archaeological Sciences, University of Bradford.

Bell M., 1983, 'Valley sediments as evidence of prehistoric land-use on the South Downs', *Proceedings of the Prehistoric Society* **49**, 118-150.

Chadwick P., 1988, 'Mineral Operator's Code of Practice - The saga continues', *Rescue News* **44**, 7.

Clarke M.R., Booth S.J., Cannell B., Carruthers R.M. and Crofts R.G., 1982, *The resource assessment of scattered drift deposits - a feasibility study in the Redditch-Solihull area*, Mineral assessment Report, Institute of Geological Sciences **115**, 97pp.

Clay P., 1986, 'Castle Donington', *Current Archaeology* **102**, ix(7), 208-211.

Cleere H.F. and Marchant P.A. (Eds.), 1988 (Sept), *British Archaeological News* (Council for British Archaeology) **3**(7), 65.

Collins G.H., 1979, 'Chalk Spindle Whorls from Buckqhoy, Orkney', Appendix to: A. Ritchie [1976-1977] Excavation of Pictish and Viking age farmsteads at Buckqhoy, Orkney, *Proceedings of the Society of Antiquaries, Scotland* **108**, 174-227.

Cornwell J.D., 1985, 'Applications of geophysical methods to mapping unconsolidated sediments in East Anglia', *Modern Geology* **9**, 187-205.

Fleming M.J. and Redknap M., 1984 (July), 'The Goodwin Sands Survey, 1983', *The Hydrographic Journal* **33**, 45-48.

Fletcher T.P., 1981, ' Geology of the Ancholme Valley', In S. McGrail (Ed.) *The Brigg 'Raft' and her prehistoric environment*, BAR British Series **89**: National Maritime Museum, Greenwich, Archaeological Series 6, 189-195.

Gostelow P., 1988, 'Engineering geology and ancient monuments in Scotland: the use of terrestrial photogrammetry as a survey method at Dunbar Castle, East Lothian', In P.G. Marinos and G.C. Koukis (eds.) *Engineering geology of ancient works, monuments and historical sites*, Balkema, Rotterdam.

Hamblin R.J.O. and Crofts R.G., 1984, 'Multispectral survey of an area of scattered drift deposits between Warwick and Redditch', *International Journal of Remote Sensing* **5**(4), 729-732.

Hull E., 1876, 'A scheme of water supply for villages, hamlets and country parishes of the central and eastern counties', *Quarterly Journal of Science* **vi**, 304-317.

Hutchinson J.N. and Gostelow T.P., 1976, 'The development of an abandoned cliff in London Clay at Hadleigh, Essex', *Philosophical Transactions of the Royal Society*, London A **283**, 557-604.

Joyce C., 1985 (October), 'Archaeologists map wrecks beneath the sea', *New Scientist*, **36**.

Leicestershire Museums, Arts & Records Service Unit, 1989, *Iron Age farmstead project at Donington le Heath Manor House*, (Unpublished pamphlet -Leicestershire Museums, Arts and Records Service).

Lynn D., 1986, *The use of thermal infrared data for the discrimination of soil and surface materials*, PhD Thesis, University of Reading.

Mathers S.J. and Zalasiewicz J.A., 1984, 'A new approach to drift mapping in East Anglia', *Quaternary Newsletter* **43**, 13-17.

Mathers S.J. and Zalasiewicz J.A., 1985(Jan), 'Defining waste and overburden associated with sand and gravel - a geophysical technique', *Quarry Management*, 27-29.

Mellars P., 1988, *Research priorities in archaeological science*, Report for Council of British Archaeology.

Pattison J. and Williamson I.T., 1986, 'The saltern mounds of north-east Lincolnshire', *Proceedings of the Yorkshire Geological Society* **46**(1), 77-83.

Plouviez J., 1988, 'Current funding and structure in British archaeology: a preliminary report', *Rescue News* **44**, 8.

Pollard M. (Ed.), 1988 (Dec), 'The environmental interface between SBAC and NERC', *Science-based Archaeology Newsletter* **1**, 2-3.

Science and Engineering Research Council, 1985, *The funding of research in science-based archaeology in universities and polytechnics: Report of a Panel chaired by Professor M.Hart, FRS*, SERC - Swindon, Wiltshire.

Thomas J., Thibodeaux L.J., Ramenofsky A.F., Field S.P., Miller B.J. and Whitmer A.M., 1988 (May), 'Archaeological Chemistry - the long-term migration of metals from archaeological contexts affects the movement of hazardous wastes', *Environmental Science and Technology* **22**(5), 480-487.

Worssam B.C. and Old R.A., 1988, 'Geology of the country around Coalville', *Memoir British Geological Survey*, Sheet **155**, 128-130.

Zalasiewicz J.A. and Mathers S.J., 1985, 'A new approach to mapping tills', *Proceedings of Engineering Techniques: report for March 1985 Conference - TILLS (Edinburgh)*, 1-5.

Zalasiewicz J.A. and Mathers S.J., 1986, 'Lithostratigraphy of the Red Norwich Crags of the Alderbury-Orford area, south-east Suffolk', *Geological Magazine* **122**(3), 287-296.

Zalasiewicz J.A., Mathers S.J. and Cornwell J.D., 1985, 'The application of ground conductivity measurements to geological mapping', *Quarterly Journal of Engineering Geology (London)* **18**, 139-118.

The Shapwick Project, Somerset: A study in need of remote sensing.

Michael Aston

Department for Continuing Education, University of Bristol

Abstract

The Shapwick Landscape Archaeology Project was set up in 1988 and is based in the Continuing Education Department of Bristol University. The aims are to examine the development of a central Somerset parish, the methodologies employed, and to involve the local people as much as is practical. The area has been extensively studied in the past by members of the Somerset Levels Project and its prehistory, particularly the Sweet track, is well known. The current project aims to try to find out when and how the present village and its medieval fields were founded. While a range of conventional archaeological and historical methods are being employed, including ground and air survey, map and document analysis and botanical and geological surveys, it is felt that real progress in identifying earlier areas of settlement will only be possible with geophysical surveys. This article argues that such survey methods will become increasingly important in landscape surveys of this type in the future.

Introduction

This note is about the Shapwick Landscape Archaeology Project in Somerset, a multi-disciplinary study of a parish. The project is intended, along with many other aspects, to examine the hypothesis that there was a change in settlement pattern from the Saxon to medieval periods from one of dispersed farmsteads to a nucleated village. It is emphasised that this aim will only be possible with the use of remote sensing techniques as other archaeological data are likely to be poorly represented or entirely lacking and conventional fieldworking techniques in pastoral landscapes on largely aceramic sites have not proved very productive in the past.

The Project

Shapwick is a medium-sized parish lying in the centre of Somerset roughly midway between Bridgwater and Glastonbury on the north side of the Polden Hills, a low east-west ridge of liassic limestone (Figure 73). The parish was part of the extensive estates of the abbey of Glastonbury for many centuries before the dissolution of the monasteries in 1539. As such it is particularly well-documented with numerous surveys and rentals from the twelfth century onwards (Postan 1973). The area has also been the focus of a great deal of archaeological research over the last two decades particularly by the Somerset Levels Project (Coles and Coles 1986). This has shown how extensively this part of Somerset was occupied in the prehistoric period with settlements and structures dating from the fifth millenium BC. The most famous of these is the Sweet Track, half of which runs through Shapwick parish (Coles, Orme and Rouillard 1984).

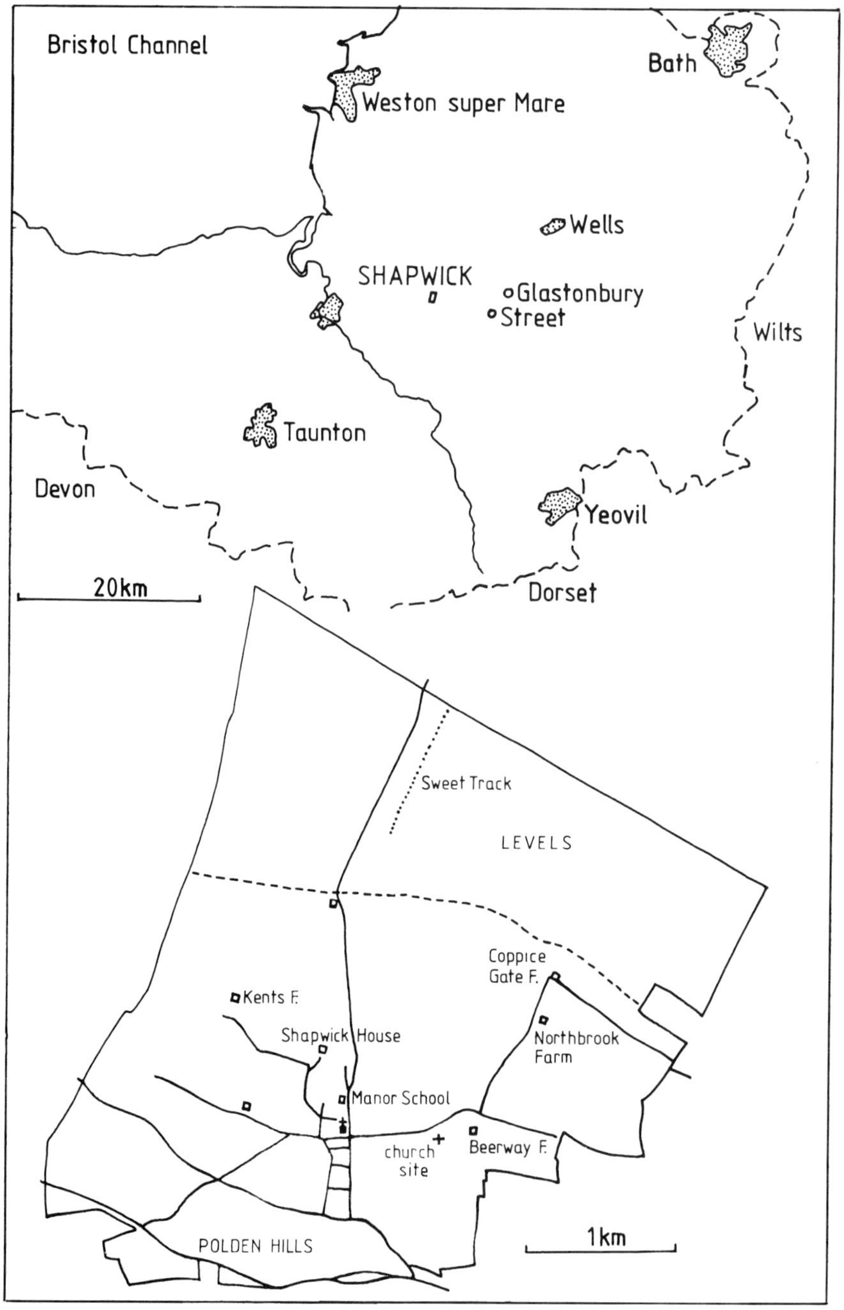

Figure 73 Shapwick parish: location and main features referred to in the text.

The writer was looking for a suitable place on which to base a project to test the above hypothesis of the nucleation of settlement. Attention was drawn to the parish of Shapwick in a study by Nicholas Corcos (Corcos 1982). He suggested, as part of his research into the enclosure of the medieval field system, that the present village of Shapwick was probably planned and that its two field system in the middle ages was so regular in layout that it too was probably planned. He also suggested that before the village there may have been a number of individual farmsteads employing a different sort of field system (Corcos 1983). Shapwick therefore seemed to provide an example of a parish in which it might be possible to show that a planned village with a regular medieval field system had replaced a more scattered pattern of single farmsteads (Aston 1988). It was hoped that by employing a battery of field and documentary techniques a definite statement could be made as to whether such a change had taken place in at least one Somerset parish. Shapwick seemed to be a good candidate to pick based on the work that had been done so far. Initially, due to existing heavy commitments of some of the likely participants, a five year project was envisaged (Aston 1989a).

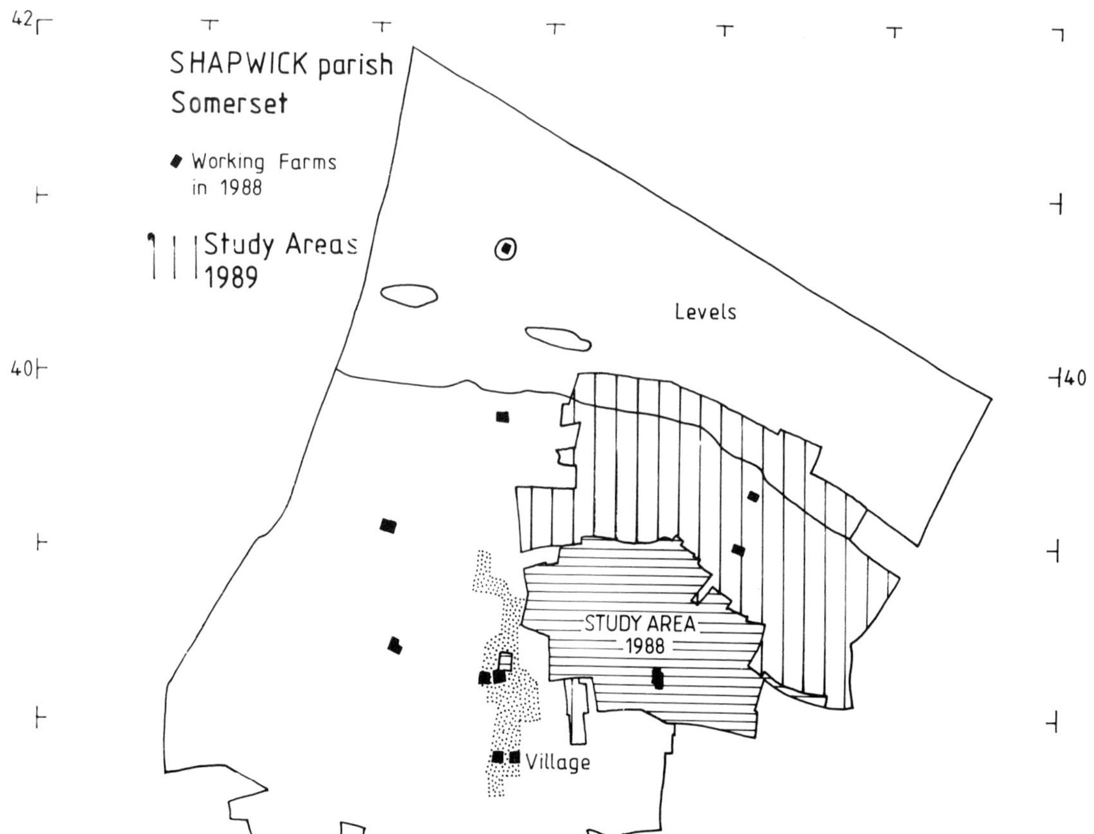

Figure 74 *Shapwick parish: areas of study in 1988 and 1989.*

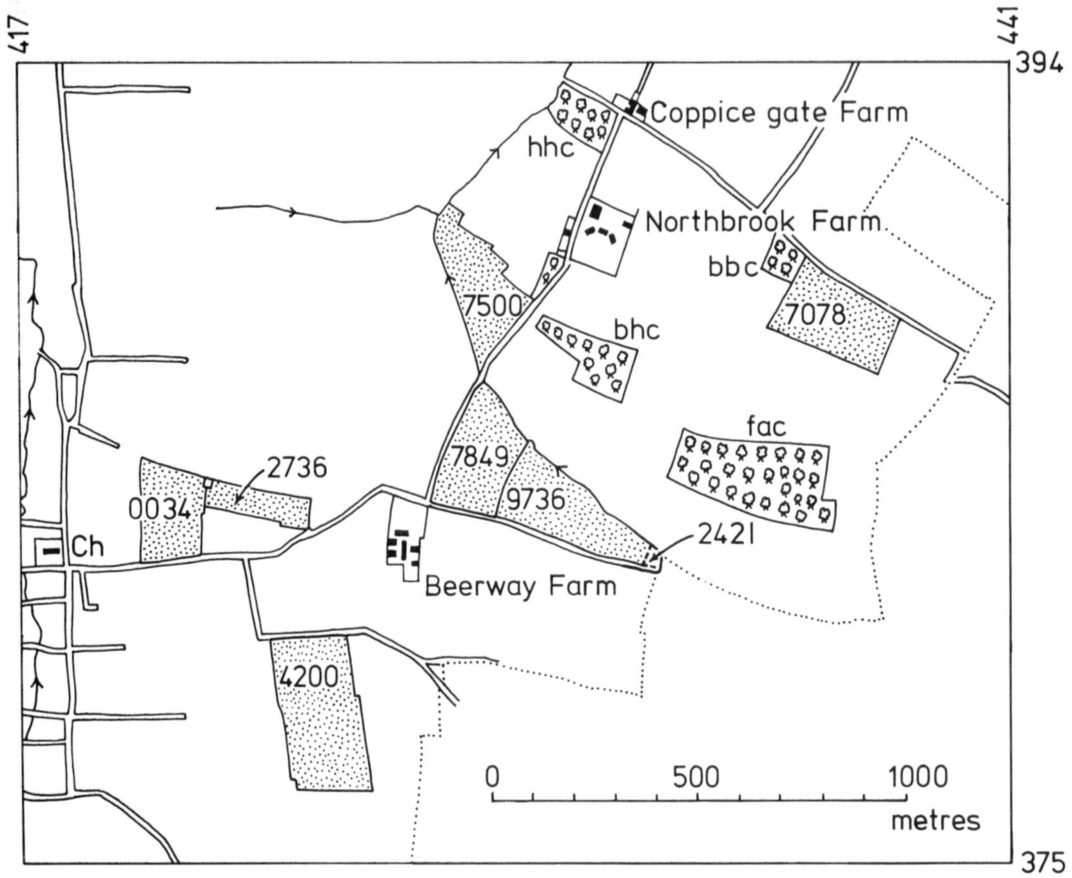

***Figure 75** Shapwick: areas of fieldwalking in 1989.*

The main landowner and several tenant farmers were contacted and work began in October 1988 with fieldwalking on the land attached to Beerway and Church Farms in the east of the parish (Figure 74). Each field was walked so that the 1971 OS 25" map base could be corrected; many of the field boundaries had been removed or altered in recent years. The current land use was also ascertained based on arable or pasture at the time of visiting. From the start of 1989 the land of two further farms was walked, Coppice Gate Farm and Northbrook Farm. In some cases this survey work drew attention to vague areas of earthworks and elsewhere to thin scatters of flints, pottery and so on which would merit attention in future; none of the material was picked up at this stage.

In June 1989 the first systematic fieldwork recording and collecting was undertaken in a field which was ploughed late and seeded with maize. In September and October further areas were systematically fieldwalked (Figure 75). Because of the limited time at the disposal of the writer and the lack of continuity of fieldwalking from one day to another, a very simple method of gridding and walking had to be applied. Having selected a field for examination based on the farmwork in hand a grid of ten by ten metre squares was laid out with reference to field boundaries on the modern OS 25" maps. Generally speaking the fields selected had been recently ploughed and most, but unfortunately not all, had been rolled and power harrowed. The weather of the Autumn of 1989 was very dry and sunny so the conditions for walking were frequently not ideal with dusty and dry solid soil. With the lack of continuity from field to field it was decided not to base the squares on the national grid but to lay out squares with reference to field boundaries.

From the outset, in consultation with Stephen Minnitt of the Somerset County Museum at Taunton, where the finds were to be deposited, it was decided to collect everything not obviously soil or bedrock. This has meant, of course, large quantities of brick, tile and modern debris as well as the earlier material. It remains to be seen if this was a wise policy in view of the vast amount of material collected but it was felt that recent activities of manuring and rubbish disposal should be of as much interest as earlier activity. A generally willing labour supply was provided by boys of the nearby Shapwick Manor School and extra-mural students. Each square was walked to a prescribed and recorded time, generally twenty minutes or half an hour.

In the remarkably dry conditions of the summer of 1989 an aerial survey was undertaken to view the extensive cropmarks which had appeared on the lias limestone as a result of the drought. This produced numerous features a lot of which are probably natural and the result of a surprisingly varied geological base and others the result of modern agricultural activity (Figures 76 and 77). Others however probably indicate earlier human activity and in two areas there are particularly interesting features. In one area on the western side of the parish 'splodges' and lines where Roman pottery is reputed to have been found and in the other, adjacent to Beerway Farm, clear parchmarks of rectangular buildings. It is known that a former church stood hereabouts; it was demolished in the early 1330's and replaced by the present 14th century church built in the middle of the village, several hundred metres away. The parchmarks seem to indicate a simple church of nave, chancel and central tower, very much like the plan of the 14th century church that replaced it. Nearby is another rectangular building, perhaps a priests house, and a round stone mound. All these features stand in an embanked enclosure next to a dried up spring.

At the same time as this archaeological work was being undertaken an extensive reassessment of the documentary sources was begun by Michael Costen. This involved the transcription of unprinted sources and a re-examination of those already printed. Eventually all the names of parcels, furlongs, tenants and so on will be computer analysed so that a pedigree for each parcel of land can be built up. From the other end of the time scale a retrogressive analysis of the development of the landscape was begun using the abundant map sources. These included the 1880s, 1906 and 1971 OS maps, the tithe map of 1839 and a series of maps of the 1750s and 1760s of both the parish and the village preserved at the Somerset County Record Office in Taunton (Aston 1989a). This analysis will enable a detailed topographical history of each plot of land to be built up for the last two centuries and it should prove possible to link this easily with the earlier abundant documentary sources.

In early 1989 a detailed buildings survey was begun, not only for the farmsteads whose fields were being studied but also of all the smaller cottages and houses of the village. This is being conducted by members of the Somerset and South Avon Vernacular Buildings Research Group who are well known in Somerset for their thorough research and prompt publication (SSAVBRG 1982-1988). As well as Shapwick House and Shapwick Manor, both of the early post-medieval period, there are several other grand houses in the village, Lawn House and the Vicarage. Outside the village, Kents and Northbrook Farms appear to be the earliest in date, the latter a fine brick Georgian-style house with flanking agricultural buildings (Figure 78). There are two dovecotes in the village and there were formerly, until the late 18th century, two large barns as well.

In the autumn of 1989 a detailed field by field geological survey was begun, conducted by Dr Peter Hardy, to provide data that could be compared with the known history of each field.

But what of the location of postulated pre-village settlements and fields? A great deal of research in eastern England has located scatters of pottery of middle and late Saxon date in a landscape which is intensively ploughed and farmed. Frequently earlier patterns of prehistoric and Roman settlements are also detected and often these sites are sealed beneath areas of medieval ridge and furrow; the earlier settlements are clearly earlier than the medieval nucleated settlements and the field systems that went with them (Taylor 1983, Hall 1981, Foard 1978 and Ford 1979). This research taken together with the study of present settlement plans has led to the development of a hypothesis on the origins of medieval rural settlement. Simply, this suggests that a predominantly dispersed pattern of farmsteads in the Anglo-Saxon period had been replaced in many areas by the early middle ages with larger more nucleated, and in many cases, planned villages (Aston 1985a and b, Aston 1989b).

In Somerset a number of factors have so far precluded the developments outlined above from being demonstrated and, although this author thought there was enough circumstantial evidence some years ago (Aston 1985b), such a change in settlement pattern in a medieval parish has yet to be demonstrated from field evidence and, elsewhere, conclusively proved by excavation. Despite a rich archaeological heritage, and considerable fieldwork on deserted medieval settlements, there is no solid tradition of fieldwalking ploughed fields in the county of Somerset. This is largely a reflection of the terrain, with much rolling pasture land, but there are archaeological considerations as well.

The need for remote sensing

Somerset is a predominantly pastoral county with large areas laid down to permanent grass; in this respect it a very different type of country for the fieldwalker to work in than eastern England with its vast areas of annually ploughed land. Those grassland areas of the county which are ploughed tend to be long leys with perhaps intervals of five to seven years between ploughings. The medieval, and possibly Saxon, church site at Beerway Farm mentioned above (Figure 77) was ploughed last in the spring of 1989, before the project was really underway; it will not be ploughed again for five years or so. In the absence of surface clues what is the fieldworker to do with such unploughed areas?

Traditionally this is where remote sensing has been applied, particularly with the use of magnetometers and resistivity meters, and where differential plant growth in the form of parchmarks and cropmarks sometimes proves useful. But so far, because of the expense of buying or hiring such equipment and the current predominant role of such geophysical techniques mainly to provide data for rescue archaeological projects where an area is to be destroyed, only very carefully selected sites have usually been examined. To my knowledge such techniques have <u>never</u> been applied to whole landscapes in this country whether or not anything or archaeological interest is suspected there (but see Carr 1982 for work in the USA).

More dramatically, archaeologists working on the Roadford project in Devon, in an almost entirely pastoral landscape, have decided to have areas ploughed which have been selected specifically for archaeological research reasons in order to retrieve scatters of pottery and artefacts. There the area is to be flooded under a reservoir and there are no farmers left in the valley to make such decisions on agricultural grounds (pers. comm. Christopher Henderson). It may be that with good relations with landowners and farmers other projects might actually persuade friendly farmers to plough up areas for the purposes of archaeological research!

Apart from the availability of ploughed land for fieldwalking however there is a greater archaeological problem. Despite an extremely rich post-Roman and Anglo-Saxon archaeology in the county of Somerset it is extremely difficult to recognise these periods from fieldwalking results. This is principally because of the lack of clearly identifiable or closely datable pottery. There is very little between the fifth century pottery of such sites as Ilchester and the tenth/eleventh century wares from such sites as Cheddar. Indeed that which is known comes from what could be considered special 'high status' sites and hence may be of little use in our attempts to identify more mundane farming settlements. So with a lack of ploughed land and a dearth of pre-conquest pottery how are we to identify the pre-village settlements in a parish like Shapwick?

One possibility lies in the documentary sources which often refer to the names of small parcels of land and areas of the common fields such as the individual furlongs. In a parish of 1284 hectares, even though about a third are low-lying levels, there is still a large area to either await periodic ploughings or employ remote sensing techniques.

Figure 76 *Shapwick: air photograph of cropmarks in the west of the parish recorded in June 1989. Most of these are probably geological in origin although Roman pottery is reputed to have been found in the locality (M Aston).*

Figure 77 *Shapwick: air photographof cropmarks adjacent to Beerway Farm taken in June 1989. The site of the church demolished in the 1330's can be seen together with another rectangular enclosure (M Aston).*

There has to be a way of narrowing down the areas which are likely to contain the early settlements. Nicholas Corcos in his research suggested that some of the names of furlongs and parcels of land referred to in the early medieval surveys contained elements that seemed to suggest habitation, or at least a different type of land use, rather than the open field areas to which they referred. Such names include 'manycrofte', 'langenworthy', 'shortenworthy', 'purycrofte', 'langworth', 'enworthie', 'shortgoldworth', and 'worthie' (Corcos 1983). Michael Costen in his surveys of Somerset field and place names has also suggested that many names, such as 'worth', 'wick', 'huish' and 'hide', seem to indicate a dispersed pattern of farmsteads and that in many cases these were later overlain by common fields attached to nucleated villages. There is a noticeable difference between the settlement patterns of west and east Somerset which seems to strengthen this suggestion. In the west, where the settlement pattern is still predominantly one of dispersion, these habitative names are still attached to old-established farmsteads; in the east of the county most of the names are found as field names (Costen 1985 and forthcoming, Aston 1989b).

If these probable habitative names mentioned in Shapwick documents could be located even approximately in the landscape they would provide more narrowly defined areas where further field archaeological research could be concentrated. So far however it has proved difficult to pinpoint most of the names even to within 500 metres or so. It may be possible to locate one of them 'shortenworthy' on the western side of the parish. We might expect this name to become something like 'shortennery' since Somerset speech frequently seems to contract these -worthy names (examples on Exmoor include Pinkworthy to Pinkery and Sweetworthy to Sweetery). On the tithe map of 1839 there are 'ennery' names for several of the closes to the west of the village; these would be good places to begin detailed survey and particularly geophysical work.

Even though detailed documentary research on individual pieces of land will eventually locate roughly the names with habitative elements contained within them, there is of course no guarantee that such names even indicate all or even some of the early settlement sites. Similarly there will certainly be other places where the names are no clue and which must be located by other methods. It will be the purpose of the rest of this paper to discuss what other lines of enquiry might be employed to locate earlier settlement sites.

Ideally it would probably be best to carry out magnetometer and resistivity surveys over the whole area of the parish in every field and plot of land. However this is not possible either practically or in terms of the cost of employing the specialists to carry out the work, even if either or both of the methods proved applicable and useful on the type of land and subsurface conditions at Shapwick, a factor that has not yet been examined. A similar reservation must also apply to the use of ground radar, a method which in any case would seem to be perhaps of more use in more deeply stratified sites (Stove and Addyman 1989). With the sort of slight traces which are often found when pre-conquest sites are excavated it is not at all clear to this writer what we might expect to see from the results of such methods on ephemeral Anglo-Saxon agricultural settlement sites. Clearly it may prove to be more useful to employ such techniques in areas where something is suspected but the reservations about whether the traces can be recognised still apply.

Every opportunity will have to be taken to maximise the results from conventional fieldwork techniques. Thus the dry summer of 1989 meant that many fields contained cropmarks and these were photographically recorded from the air. First examination however suggests that it may be difficult in this area to distinguish natural crop/grass marks and geological features from those formed as a result of man's activities (Figure 79). And as in the case of the remote sensing methods the writer is not at all sure of what sorts of cropmarks, if any, might indicate the slight traces of pre-conquest agricultural buildings and structures.

Fieldwalking will have to be employed every time a field is ploughed. With the long ley periods between ploughings of much of the grassland, any surface disturbance clearly represents a great opportunity to see what indications there might be of earlier settlements. Here again there will be difficulties. Even in the east of England, Anglo-Saxon pottery is notoriously elusive and scarce to the extent that some fieldwalkers have resorted to examining areas on their knees! (Williamson 1984, 1985). Although everything found in fieldwalking in this project will be retained for the time being, and hence all material can be examined by specialists later on, the recognition of the earlier Anglo-Saxon and medieval pottery may prove a problem in the field leading to a failure to perhaps appreciate the significance of the material when and if it is located. This will not be a major problem so long as the recording system remains as detailed as it is at the moment under the supervision of Dr Christopher Gerrard.

Despite past antagonism between the archaeologists and metal detectorists most directors of field projects now realise that used appropriately metal detectors can be a useful tool in field survey. In the case of this project detector surveys of ploughed fields carried out at the same time as fieldwalking might be expected to produce complimentary information. With pre-conquest sites in particular single coins or relevant pieces of diagnostic metalwork are just as likely, or perhaps more likely, to turn up as pottery of the period and to indicate possible settlement sites.

One method which might prove particularly useful in locating earlier settlement sites is phosphate analysis, particularly if used alongside other survey and geophysical methods (Gurney 1985). Surprisingly such analysis has not been used widely in this country on landscape survey work or medieval settlement sites (Alexander and Roberts 1978). This would seem to be an appropriate method to use at Shapwick alongside fieldwalking, geophysical and geological surveys.

Figure 78 *Shapwick: air photograph of Northbrook Farm taken in June 1989. This farm dates from the late eighteenth century and is one of only two farmsteads built away from the village on the open fields before the nineteenth century (M Aston).*

Figure 79 *Shapwick: air photograph of geological features recorded in June 1989 (M Aston).*

With such extensive areas of permanent and non-permanent pasture one method which might prove to be of some significance is the digging of randomly selected trial holes of say two by two metres. Clearly such small holes are not going to easily indicate structures from earlier periods but the return of information on scattered finds, depth of topsoil and other layers and potential sources of environmental material might prove of significance. On the basis of 1284 hectares of land to be examined or, more realistically, c800 hectares without the levels which account for about a third, a ten percent sample of 80 hectares would clearly be impossible with the likely resources at the disposal of the project. Indeed even a one percent sample of 8 hectares represents 80,000 square metres randomly selected would entail a large and consistent labour input in the field, without any consideration of the post-digging analysis. Clearly 80,000 one square metre holes, or even 20,000 two by two metre holes is likely to be beyond the time and labour resources of most archaeologists! And yet to conduct a rigorous landscape archaeology project and, in particular, to locate the ephemeral traces of early settlements such a massive input, or something like it, is perhaps necessary. The reality of trying to do thorough fieldwork over even a small percentage of the parish in the time available, and with the people and resources at the disposal of the project, is already causing concern; the prospect of the effort required in random sampling a landscape archaeologically is mind-numbing!

Concluding remarks

This paper has mainly consisted of the authors thoughts on a project which has hardly begun. The project has very specific aims although inevitably more information will be found for other periods and more questions will be raised than will be answered as time goes on. The particular problems associated with trying to answer these specific questions have formed a large part of the discussion above. They reveal, not only the lack of knowledge, experience and expertise of the author, but also some of the difficulties in trying to find sites and landscapes of a particular period in countryside not traditionally associated with yielding such information. It is clear that traditional approaches and traditional methods will only take the investigation so far. It is equally clear that a more multidisciplinary and rigorous approach has a greater chance of success than perhaps a more open ended and pragmatic one. What emerges above all however is the very great role that remote sensing could and should have in such a project. Given the difficulties likely to be encountered in site and feature location, the predominantly grassland landscape and the paucity of archaeological material likely to be found to give any chronological indications, remote sensing offers the greatest chance to locate something, anything, that might then be subject to closer survey, sampling or excavation. Perhaps in a few years time the application of geophysical and remote sensing methods will be more standard in such projects and we will then have a chance of answering some of the questions raised by such projects as this at Shapwick.

Acknowledgements

The landowner, Lord Vestey, his land agent Mr Bill Robbins and the farmers of Shapwick, particularly David and Stan Jenkins and Mr and Mrs Graham Elliott who have readily given access to their land and been supportive of the project in more ways than I could have hoped for. I am grateful to Michael Costen and Paul Spoerry for constructively commenting on an earlier draft of this paper.

References

Alexander M.J. and Roberts B.K., 1978, 'Low Buston: a study in soil phosphate analysis' *Archaeologia Aeliana* 5th series **6,** 107-116.

Aston M., 1985a, *Interpreting the Landscape,* London Batsford.

Aston M., 1985b, 'Rural Settlement in Somerset: some preliminary thoughts' in Hooke D. (Ed.) *Medieval Villages: A Review of Current Work,* Oxford University Committee for Archaeology Monograph **5**, 81-100.

Aston M., 1988, 'Settlement Patterns and Forms' in Aston M. (Ed.) *Aspects of the Medieval Landscape of Somerset: contributions to the landscape history of the county,* Taunton Somerset County Council, 67-81.

Aston M., 1989a, *The Shapwick Project: A Topographical and Historical Study,* 1988 Report Bristol University, Extra-mural department.

Aston M.,1989b, 'The development of medieval settlement in Somerset' in Higham R. (Ed.) *Landscape and Townscape in the South West,* Exeter Studies in History **22,** University of Exeter, 19-40.

Carr C., 1982, *Handbook on soil resistivity surveying*, Central American Archaeological Press, Evanston, Illinois.

Coles J.M., Orme B.J. and Rouillard S.E. (Eds.), 1984, *Somerset Levels Papers* **10**.

Coles B. and Coles J.M., 1986, *Sweet Track to Glastonbury,* London, Thames and Hudson.

Corcos N., 1982, *Shapwick: the enclosure of a Somerset parish 1515-1839*, University of Leicester Department of English Local History, unpublished MA thesis 1981-2.

Corcos N., 1983, 'Early estates on the Poldens and the origin of settlement at Shapwick' *Somerset Archaeology and Natural History* **127**, 47-53.

Costen M., 1985, 'Rimpton in Somerset - a late Saxon Estate', *Southern History* **7**, 13-24.

Foard G., 1978, 'Systematic fieldwalking and the investigation of Saxon settlement in Northamptonshire', *World Archaeology* **9**, 357-374.

Ford W.J., 1979, 'Some settlement patterns in the central region of the Warwickshire Avon' in Sawyer P.H. (Ed.) *English Medieval Settlement,* London, Edward Arnold, 143-163.

Gurney D.A., 1985, *Phosphate Analysis of Soils: a guide for the Field Archaeologist,* Institute of Field Archaeologists Technical Paper Number **3**.

Hall D., 1981, 'The Origins of Open-field Agriculture - The Archaeological Fieldwork Evidence' in Rowley T. (Ed.) *The Origins of Open Field Agriculture,* Croom Helm, London.

Posten M.M., 1973, 'Glastonbury estates in the twelfth century' in *Essays on Medieval Agriculture and General Problems of the Medieval Economy,* Cambridge University Press, 249-277.

SSAVBRG, 1982-88, Somerset and South Avon Vernacular Building Research Group 1982-1988 *Long Load (1982), West and Middle Chinnock (1984), Alford and Lovington (1986), and Batcombe (1988)* Somerset Rural Life Museum Glastonbury.

Stove G.C. and Addyman P.V., 1989, 'Ground probing impulse radar: an experiment in archaeological remote sensing in York' *Antiquity* **63** 239 - June 1989, 337-342.

Taylor C., 1983, *Village and Farmstead: A History of Rural Settlement in England,* George Philip, London.

Williamson T., 1984, *Roman and Medieval Settlement in North-West Essex,* University of Cambridge unpublished PhD thesis.

Williamson T., 1985, 'Sites in the Landscape: Approaches to the post-Roman settlement of South-East England' *Archaeological Review from Cambridge* **4.1** - Spring 1985, 1-64.